New Data Structures and Algorithms for Logic Synthesis and Verification

Luca Gaetano Amaru

New Data Structures and Algorithms for Logic Synthesis and Verification

 Springer

Luca Gaetano Amaru
Synopsys Inc.
Santa Clara, CA
USA

ISBN 978-3-319-82753-7   ISBN 978-3-319-43174-1   (eBook)
DOI 10.1007/978-3-319-43174-1

Printed on acid-free paper

This Springer imprint is published by Springer Nature
The registered company is Springer International Publishing AG Switzerland

Success is not final, failure is not fatal:
It is the courage to continue that counts.

—Winston Churchill

To my parents

Preface

The strong interaction between *Electronic Design Automation* (EDA) tools and *Complementary Metal-Oxide Semiconductor* (CMOS) technology contributed substantially to the advancement of modern digital electronics. The continuous downscaling of CMOS *Field Effect Transistor* (FET) dimensions enabled the semiconductor industry to fabricate digital systems with higher circuit density at reduced costs. To keep pace with technology, EDA tools are challenged to handle both digital designs with growing functionality and device models of increasing complexity. Nevertheless, while the downscaling of CMOS technology requires more complex physical design models, the logic abstraction of a transistor as a switch has not changed even with the introduction of 3D FinFET technology. As a consequence, modern EDA tools are fine-tuned for CMOS technology and the underlying design methodologies are based on CMOS logic primitives, i.e., negative unate logic functions. While it is clear that CMOS logic primitives will be the ultimate building blocks for digital systems in the next 10 years, no evidence is provided that CMOS logic primitives are also the optimal basis for EDA software. In EDA, the efficiency of methods and tools is measured by different metrics such as (i) the result quality, for example the performance of an automatically synthesized digital circuit, (ii) the runtime, and (iii) the memory footprint on the host computer. With the aim to optimize these metrics, the accordance to a specific logic model is no longer important. Indeed, the key to the success of an EDA technique is the expressive power of the logic primitives' handling and solving the problem, which determines the capability to reach better metrics.

In this book, we investigate new logic primitives for electronic design automation tools.

We improve the efficiency of logic representation, manipulation, and optimization tasks by taking advantage of majority and biconditional logic primitives. We develop synthesis tools exploiting the majority and biconditional logic expressiveness. Our tools show strong results as compared to state-of-the-art academic and commercial synthesis tools. Indeed, we produce the best (public) results for many circuits in combinational benchmark suites. On top of the enhanced synthesis power, our methods are also the natural and native logic abstraction for circuit design in emerging nanotechnologies, where majority and biconditional logic are the primitive gates for physical implementation.

We accelerate formal methods by (i) studying core properties of logic circuits and (ii) developing new frameworks for logic reasoning engines. Thanks to the majority logic representation theory, we prove nontrivial dualities for the property checking problem of logic circuits. Our findings enable sensible speedups in solving circuit satisfiability. With the aim of exploiting further the expressive power of majority logic, we develop an alternative Boolean satisfiability framework based on majority functions. We prove that the general problem is still intractable, but we show practical restrictions that instead can be solved efficiently. Finally, we focus on the important field of reversible logic where we propose a new approach to solve the equivalence checking problem. We define a new type of reversible miter over which the equivalence test is performed. Also, we represent the core checking problem in terms of biconditional logic. This enables a much more compact formulation of the problem as compared to the state of the art. Indeed, it translates into more than one order of magnitude speedup for the equivalence checking task, as compared to the state-of-the-art solution.

We argue that new approaches to solve core EDA problems are necessary, as we have reached a point in technology where keeping pace with design goals is tougher than ever.

Santa Clara, USA   Luca Gaetano Amaru

Acknowledgments

I would like to express my sincere gratitude to my advisor Prof. Giovanni De Micheli for giving me the opportunity to pursue doctoral studies within the Integrated Systems Laboratory (LSI) at EPFL. His guidance, motivation, and immense knowledge helped me in all time of research. I could not have imagined having a better advisor and mentor for my Ph.D. studies. I am thankful to my co-advisor, Prof. Andreas Burg, for his valuable advice and encouragement.

Besides my advisors, I would like to thank Dr. Pierre-Emmanuel Gaillardon, who provided me tremendous support and guidance through my doctoral studies. Without his precious help this book would not have been possible.

Furthermore, I am very grateful to Prof. Subhasish Mitra for the inspiring discussions on the interaction between EDA and nanotechnologies and for giving me the opportunity to be a visiting student at Stanford University.

I would like to express my deepest appreciation to Prof. Maciej Ciesielski, Dr. Alan Mishchenko, Prof. Anupam Chattopadhyay, Dr. Robert Wille, and Dr. Mathias Soeken for the great research collaborations we had, which are also part of this book.

My sincere thanks also go to Profs. Paolo Ienne, Joseph Sifakis, Subhasish Mitra, and Enrico Macii for their insightful comments on this work.

Last but not least, I would like to thank my parents for their unconditional love and continuous support.

Contents

Chapter 1
Introduction

The strong interaction between *Electronic Design Automation* (EDA) tools and *Complementary Metal-Oxide Semiconductor* (CMOS) technology contributed substantially to the advancement of modern digital electronics. The continuous downscaling of CMOS *Field Effect Transistor* (FET) dimensions enabled the semiconductor industry to fabricate digital systems with higher circuit density and performance at reduced costs [1]. To keep pace with technology, EDA tools are challenged to handle both digital designs with growing functionality and device models of increasing complexity. Nevertheless, whereas the downscaling of CMOS technology is requiring more complex physical design models, the logic abstraction of a transistor as a switch has not changed even with the introduction of 3D FinFET technology [2]. As a consequence, modern EDA tools are fine tuned for CMOS technology and the underlying design methodologies are based on CMOS logic primitives, i.e., negative unate logic functions. While it is clear that CMOS logic primitives will be the ultimate building blocks for digital systems in the next ten years [3], no evidence is provided that CMOS logic primitives are also the optimal basis for EDA software. In EDA, the efficiency of methods and tools is measured by different metrics such as (i) the result quality, for example the performance of an automatically synthesized digital circuit, (ii) the runtime and (iii) the memory footprint on the host computer. With the aim to optimize these metrics, the accordance to a specific logic model is no longer important. Indeed, the key to the success of an EDA technique is the expressive power of the logic primitives handling and solving the problem, which determines the capability to reach better metrics.

Overall, this book addresses the general question: *"Can EDA logic tools produce better results if based on new, different, logic primitives?"*. We show that the answer to this question is affirmative and we give pragmatic examples. We argue that new approaches to solve core EDA problems are necessary, as we have reached a point of technology where keeping pace with design goals is tougher than ever.

© Springer International Publishing Switzerland 2017
L.G. Amaru, *New Data Structures and Algorithms for Logic Synthesis and Verification*, DOI 10.1007/978-3-319-43174-1_1

1.1 Electronic Design Automation

EDA is an engineering domain consisting of algorithms, methods and tools used to design complex electronic systems. Starting from a *high-level description* of an electronic system, a typical EDA flow operates on several logic abstractions and produces a final implementation in terms of primitive technology components [4]. When targeting an *Application Specific Integrated Circuit* (ASIC) technology, the final product is a GDSII file, which represents planar geometric shapes ready for photomask plotting and successive fabrication [5]. When targeting a *Field-Programmable Gate Arrays* (FPGAs) technology, the final product is a binary file, which is used to (re)configure the FPGA device [6].

The main steps involved in the design flow are *high-level synthesis*, *logic synthesis* and *physical design*, also called low level synthesis, which consists of *placement and routing* [4]. They are depicted by Fig. 1.1. *High-level synthesis* converts a *programming language description* (or alike) of a logic system into a *Register-Transfer Level* (RTL) netlist. *Logic synthesis* optimizes and maps a logic circuit, from an RTL specification, onto standard cells (ASICs) or look-up tables (FPGAs). *Placement* assigns physical resources to the mapped logic elements, i.e., standard cells inside a chip's core area (ASICs) or programmable logic blocks (FPGAs). *Routing* interconnects the placed logic elements, i.e., sets wires to properly connect the placed standard cells (ASICs) or creates routing paths between programmable logic elements in a reconfigurable device (FPGAs). All these three steps are subject to area, delay and power minimization metrics. Nowadays, the clear separation between design steps fade away in favor of an integrated approach better dealing with design closure [7]. Contemporary design techniques are fine tuned for CMOS technology. For example, most logic synthesis data structures and algorithms are based on CMOS logic primitives, e.g., negative unate logic functions [4]. Placement and routing algorithms matured with the technological evolution of CMOS down to the nano-scale [3]. Logic or physical characteristics of CMOS technology have been strong progress drivers for modern design flows.

In parallel to the synthesis flow, verification techniques check that the designed system conforms to specification [8]. *Simulation* and *formal methods* are two

Fig. 1.1 Design flow

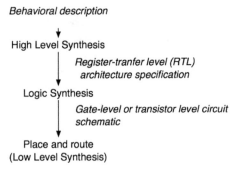

Fig. 1.2 Design verification
methods

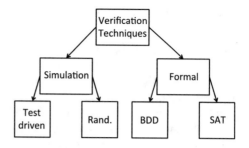

popular verification approaches [8]. Simulation techniques compute the output values for given input patterns using simulation models [9]. If the output values mismatch the given specification then verification fails. Simulation-based verification formally succeeds only if the output values match the specification for all input patterns. Because of the exponential space of input patterns, it is impractical to verify overall designs by simulations. Nevertheless, random simulation techniques are still used as fast bugs hunters. When an exact answer is needed, formal methods precisely prove whether the system conforms to specification or not. In formal methods, specification and design are translated into mathematical models [8]. Formal verification techniques prove correctness with various sorts of mathematical reasoning. It explores all possible cases in the generated mathematical models. Popular mathematical models used in formal methods include mainly Boolean functions/expressions, first order logic, and others. The main reasoning engines used are binary decision diagrams [10] and satisfiability methods [11]. Figure 1.2 depicts the aforementioned verification environment by means of a diagram.

In this book, we focus on the *logic synthesis* and *formal methods* sub-fields of EDA.

1.2 Modern EDA Tools and Their Logic Primitives

Modern EDA tools operate on logic abstractions of an electronic system. These logic abstractions are based on some primitive logic operators over which the synthesis and verification processes are performed. The expressive power and manipulation properties of the logic primitives employed ultimately determine the quality of the EDA tasks accomplished. We review hereafter the basic logic primitives driving logic synthesis and formal verification tools.

1.2.1 Logic Synthesis

In logic synthesis, the main abstraction is a logic circuit, also called logic network, which is defined over a set of primitive logic gates. Very popular primitive gates in logic synthesis are AND, OR and INV. While there are expensive (in terms of runtime) synthesis techniques operating on truth tables and global functions, most practical synthesis methods exploit the local functionality of primitive gates over which the circuit itself is described. For example, two-level AND-OR logic circuits, also called *Sum-Of-Products* (SOPs), are synthesized by manipulating cubes and their sum [12]. As cubes are inherently AND functions and their sum is inherently an OR function, two-level logic synthesis is based on AND/OR logic primitives [12]. Another example is about multi-level logic circuits and their synthesis [13]. In multi-level logic representations, logic gates may have an unbounded functionality, meaning that each element can represent an arbitrary logic function. However, these logic elements are often represented internally as SOP polynomials which are factorized into AND/ORs via algebraic methods [13]. Therefore, also multi-level logic synthesis operates on AND/OR logic primitives [13].

1.2.2 Formal Methods

In formal methods, the main logic abstraction is a formal specification. A formal specification can be a logic circuit, a Boolean formula or any other formal language capable of exhaustively describing the property under test. Ultimately, a formal specification is translated into a mathematical logic formula. To prove properties of the formal specification, two core reasoning engines are very popular in formal methods: binary decision diagrams [10] and Boolean satisfiability [11]. Binary decision diagrams are a data structure to represent Boolean functions. They are driven by the Shannon's expansion to recursively decompose a Boolean function into cofactors until the constant logic values are encountered. Reduced and ordered binary decision diagrams are unique for a given variable order, i.e., canonical. This feature enables efficient property checking. From a logic circuit perspective, the Shannon's expansion is equivalent to a 2:1 multiplexer (MUX), which therefore is the logic primitive driving binary decision diagrams [10]. Boolean satisfiability consists of determining whether there exists or not an assignment of variables so that a Boolean formula evaluates to true. The standard data structure supporting Boolean satisfiability is the *Conjunctive Normal Form* (CNF), which is a conjunction (AND) of clauses (OR). In other words, this data structure is a two-level OR-AND logic circuits, also called a *Product of Sums* (POS). The CNF satisfiability problem is solved through reasoning on clauses (ORs) and how they interact via the top conjunction operator (AND). It follows that standard satisfiability techniques are based on OR/AND logic primitives [11].

1.3 Research Motivation

Nowadays, EDA tools face challenges tougher than ever. On the one hand, design sizes and goals in modern CMOS technology approach the frontier of what is possibly achievable. On the other hand, post-CMOS technologies bring new computational paradigms for which standard EDA tools are not suitable. New research in fundamental EDA tasks, such as logic synthesis and formal verification, is key to handle this situation.

1.3.1 Impact on Modern CMOS Technology

Present-day EDA tools are based on CMOS logic primitives. For example, AND/OR logic functions, which are the basis for series/parallel gate design rules, drive several synthesis techniques. Similarly, MUX logic functions, which are the primitives for CMOS pass-transistor logic, are the building blocks for canonical data structures. While there is no doubt that these primitives will be the physical building blocks for CMOS digital systems in the next ten years [3], the use of new, more expressive, logic primitives in design and verification methods can improve the computational power of EDA tools.

Indeed, the study of new logic primitives can extend the capabilities of logic synthesis and formal verification tools already in CMOS technology. Exploiting new logic primitives, synthesis tools can reach points in the design space not accessible before [14]. Formal methods based on different logic primitives can solve faster an important class of problems, e.g., the verification of arithmetic logic [15], the verification of reversible logic [16], etc.

1.3.2 Impact on Beyond CMOS Technologies

Considering instead post-CMOS technologies, studying new logic primitives is necessary because many emerging nanotechnologies offer an enhanced functionality over standard FET switches [17].

For example, double-gate silicon nanowire FETs [18], carbon nanotube FETs [19], graphene FETs [20, 21] and organic FETs [22] can be engineered to allow device polarity control. The switching function of these devices is biconditional on both gates (polarity and control) values. Four-terminals and six-terminals nanorelays in [23, 24], respectively, operate similarly. The source to drain connection in these nanorelays is controlled by the gate to body voltage sign and amplitude. In the binary domain, this corresponds to a *bit comparator* between the gate and body logic values. Also reversible logic gates, such as Toffoli gates, embed the biconditional connective in their operation [25]. Indeed, biconditional (XNOR) operations are easily reversible

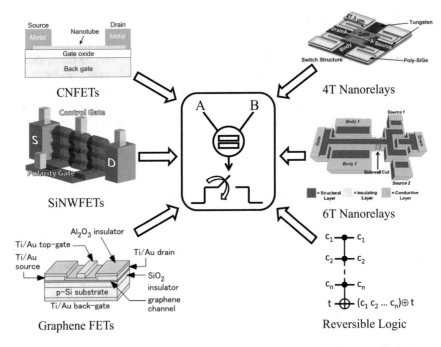

Fig. 1.3 Common logic abstraction for SiNWFETs, CNFETs, graphene FETs, reversible logic and nanorelays. Logic model: switch driven by a comparator

while other logic operations, such as conjunctions and disjunctions, are not. All these devices operate as a switch driven by a *single bit comparator*. Figure 1.3 depicts the common logic abstraction for those comparator-intrinsic nanodevices.

Other promising nanodevices, such as *Spin-Wave Devices* (SWD) [26–28], *Resistive RAM* (RRAM) [29, 30] and graphene reconfigurable gates [31], operate using different physical phenomena than standard FETs. For example, SWD uses spin waves as information carrier while CRS logic behavior depends on the previous memory state. In those nanotechnologies, the circuit primitive is not anymore a standard switch but a three-input majority voter. Note that there are other nanotechnologies where majority voters are the circuit primitive. Quantum-dot cellular automata is one well-known voting-intrinsic nanotechnology [32]. Also, DNA strand displacement recently showed the capability to implement voting logic [33]. Figure 1.4 depicts the common logic abstraction for these voting-intrinsic nanodevices.

In this context, EDA tools capable of natively handling such enhanced functionality are essential to permit a fair evaluation on nanotechnologies with logic abstractions different than standard CMOS [34].

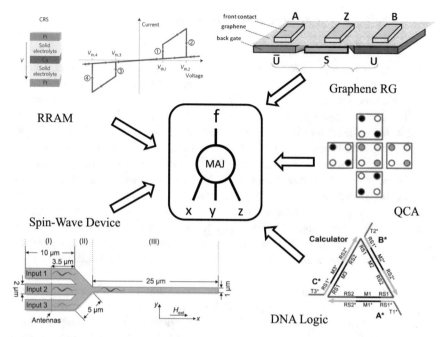

Fig. 1.4 Common logic abstraction for SWD, RRAM, Graphene reconfigurable gates, QCA and DNA logic. Logic model: majority voter

1.4 Contributions and Position with Respect to Previous Work

This book is centered around logic synthesis and formal methods. For the sake of completeness, we also include our results in the area of nanotechnology design. Our contributions can be classified into two main categories.

1.4.1 Logic Representation, Manipulation and Optimization

1.4.1.1 Contributions

We develop new compact representations for logic functions, together with powerful manipulation and optimization techniques. The two main topics here are *biconditional logic* and *majority logic*.

1.4.1.2   Position with Respect to Previous Work

Regarding *logic representation, manipulation and optimization*, state-of-the-art design tools make extensive use of homogeneous logic networks. Homogeneous logic networks are directed acyclic graphs where all internal nodes represent the same logic function and edges are possibly complemented in order to preserve universality. *And-Inverter Graphs* (AIGs) are homogeneous logic networks driven by the AND logic function [35]. AIGs are widely used in logic optimization. AIG optimization algorithms are typically based on fast and local rewriting rules, together with traditional Boolean techniques [35–37]. *Binary Decision Diagrams* (BDDs) are homogeneous logic networks driven by the MUX logic function [10]. With specific ordering and reduction rules, BDDs are canonical, i.e., unique for a logic function and variable order [10]. BDDs are commonly employed both as a representation structure and as a logic manipulation engine for optimization. Indeed, the canonicity of BDDs enables efficient computation of cofactors, Boolean difference and approximation of *don't care* sets, all important features to logic optimization techniques [38, 39].

Our contributions in this category focus on homogeneous logic networks as well. We propose *Majority-Inverter Graphs* (MIGs), an homogeneous logic network driven by ternary majority logic functions. As majority functions can be configured to behave as AND/ORs, MIGs can be more compact than AIGs. Moreover, MIG manipulation is supported by a sound and complete algebraic framework and unique Boolean properties. Such features makes MIG optimization extremely effective as compared to the state-of-the-art counterparts. We propose *Biconditional Binary Decision Diagrams* (BBDDs), a BDD-like homogeneous logic network where branching decisions are biconditional on two variables per time rather than on only one. From a theoretical perspective, considering two variables per time enhances the expressive power of a decision diagram. Nevertheless, BBDDs are still canonical with respect to specific ordering and reduction rules. BBDDs improve the efficiency of traditional EDA tasks based on decision diagrams, especially for arithmetic intensive designs. Indeed, BBDDs are smaller than BDDs for notable arithmetic functions, such as binary addition and majority voting. On the other hand, BBDDs represent the natural and native design abstraction for emerging technologies where the circuit primitive is a comparator, rather than a simple switch.

1.4.2   Boolean Satisfiability and Equivalence Checking

1.4.2.1   Contributions

We study logic transformations to speed up satisfiability check in logic circuits. We develop an alternative Boolean satisfiability framework based on majority logic rather than standard conjunctive normal form. Finally, we propose a new approach to solve sensibly faster the combinational equivalence checking problem for reversible logic.

1.4.2.2 Position with Respect to Previous Work

For the Boolean *SATisfiability* (SAT) problem, the state-of-the-art solution use a *Conjunctive Normal Form* (CNF) formulation solved by modern variants of the *Davis Putnam Logemann Loveland* (DPLL) algorithm, such as conflict-driven clause learning [11]. For the *Combinational Equivalence Checking* (CEC) problem, the state-of-the-art solution first creates a miter circuit by XOR-ing bit-wise the outputs of the two circuits under test. Then, it uses simulation and BDD/SAT sweeping on the input side (i.e., proving equivalence of some internal nodes in a topological order), interleaved with attempts to run SAT on the outputs (i.e., proving equivalence of all the outputs to constant 0) [40].

Our contributions in this category focus on alternative SAT formulations and CEC solving approaches. We define a *Majority Normal Form* (MNF), a two-level logic representation form based on generic n-ary majority operators. When described over a MNF, the SAT problem has remarkable properties. For example, practical restrictions of the MNF-SAT problem can be solved in polynomial time. Considering instead circuit satisfiability, we discover circuit dualities useful to speed-up SAT solving via parallel execution. Moving to CEC, we focus on the problem of checking the equivalence of reversible circuits. Here, the state-of-the art CEC solution is still the standard *miter-sweeping-SAT* one. We propose a different type of miter, obtained by cascading the two reversible circuits under test in place of XOR-ing them. As a result, we do not aim at proving the unSAT of the outputs anymore, but we aim at proving that the outputs all represent the identity function. In this scenario, we propose an efficient XOR-CNF formulation of the identity check problem which is solvable via Gaussian elimination and SAT. Such reversible CEC flow decreases the runtime by more than one order of magnitude as compared to state-of-the-art solutions.

In both categories Sects. 1.4.1 and 1.4.2, our research contributions exploit new logic primitives to approach fundamental EDA problems from a different, unconventional, perspective.

1.5 Book Organization

This book is divided into two parts: *Logic Representation, Manipulation and Optimization* and *Boolean Satisfiability and Equivalence Checking*. For the sake of clarity and readability, each chapter comes with separate background, notations and bibliography sections.

Part 1 Logic Representation, Manipulation and Optimization. Chaps. 2 and 3.

Chapter 2 presents *Biconditional Binary Decision Diagrams* (BBDDs), a novel canonical representation form for Boolean functions. BBDDs are binary decision diagrams where the branching condition, and its associated logic expansion, is biconditional on two variables. Empowered by reduction and ordering rules, BBDDs are remarkably compact and unique for a Boolean function. BBDDs improve the efficiency of traditional EDA tasks based on decision diagrams, especially for arithmetic intensive designs. BBDDs also represent the natural and native design abstraction for emerging technologies where the circuit primitive is a comparator, rather than a simple switch. Thanks to an efficient BBDD software package implementation, we validate (1) speed-up in traditional decision diagrams and (2) improved synthesis of circuits in traditional and emerging technologies.

Chapter 3 proposes a paradigm shift in representing and optimizing logic by using only majority (MAJ) and inversion (INV) functions as basic operations. We represent logic functions by *Majority-Inverter Graph* (MIG): a directed acyclic graph consisting of three-input majority nodes and regular/complemented edges. We optimize MIGs via a new Boolean algebra, based exclusively on majority and inversion operations, that we formally axiomatize in this book. As a complement to MIG algebraic optimization, we develop powerful Boolean methods exploiting global properties of MIGs, such as bit-error masking. MIG algebraic and Boolean methods together attain very high optimization quality. Furthermore, MIG optimization improves the synthesis of emerging nanotechnologies whose logic primitive is a majority voter.

Part 2 Boolean Satisfiability and Equivalence Checking. Chaps. 4–6.

Chapter 4 establishes a non-trivial duality between tautology and contradiction check to speed up circuit satisfiability (SAT). Tautology check determines if a logic circuit is true in every possible interpretation. Analogously, contradiction check determines if a logic circuit is false in every possible interpretation. A trivial transformation of a (tautology, contradiction) check problem into a (contradiction, tautology) check problem is the inversion of all outputs in a logic circuit. In this work, we show that exact logic inversion is not necessary. We give operator switching rules that selectively exchange tautologies with contradictions, and vice versa. Our approach collapses into logic inversion just for tautology and contradiction extreme points but generates non-complementary logic circuits in the other cases. This property enables solving speed-ups when an alternative, but equisolvable, instance of a problem is easier to solve than the original one. As a case study, we investigate the impact on SAT. We show a 25 % speed-up of SAT in a concurrent execution scenario.

Chapter 5 introduces an alternative two-level logic representation form based solely on majority and complementation operators. We call it *Majority Normal Form* (MNF). MNF is universal and potentially more compact than its CNF and DNF counterparts. Indeed, MNF includes both CNF and DNF representations. We study the problem of MNF-SATisfiability (MNF-SAT) and we prove that it belongs to the NP-complete complexity class, as its CNF-SAT counterpart. However, we show practical restrictions on MNF formula whose satisfiability can be decided in polynomial time. We finally propose a simple core procedure to solve MNF-SAT, based on the intrinsic functionality of two-level majority logic.

Chapter 6 presents a new approach for checking the combinational equivalence of two reversible circuit significantly faster than the state-of-the-art. We exploit inherent characteristics of reversible computation, namely bi-directional (invertible) execution and the XOR-richness of reversible circuits. Bi-directional execution allows us to create an identity miter out of two reversible circuits to be verified, which naturally encodes the equivalence checking problem in the reversible domain. Then, the abundant presence of XOR operations in the identity miter enables an efficient problem mapping into XOR-CNF satisfiability. The resulting XOR-CNF formulas are eventually more compact than pure CNF formulas and potentially easier to solve. Experimental results show that our equivalence checking methodology is more than one order of magnitude faster, on average, than the state-of-the-art solution based on established CNF-formulation and standard SAT solvers.

Chapter 7 concludes the book. A summary of research accomplishments is presented, which affirmatively answers the question: *"Can EDA logic tools produce better results if based on new, different, logic primitives?"*. Possible future works are finally discussed.

References

1. G.E. Moore, Cramming more components onto integrated circuits. Proc. IEEE **86**(1), 82–85 (1998)
2. D. Hisamoto et al., FinFET-a self-aligned double-gate MOSFET scalable to 20 nm. IEEE Trans. Electron Devices **47**(12), 2320–2325 (2000)
3. M. Bohr, Technology Insight: 14 nm Process Technology—Opening New Horizons, Intel Developer Forum 2014, San Francisco
4. G. De Micheli, *Synthesis and Optimization of Digital Circuits* (McGraw-Hill Higher Education, United States, 1994)
5. J. Buchanan, The GDSII Stream Format, June 1996
6. S. Brown et al., *Field-Programmable Gate Arrays*, vol. 180 (Springer Science & Business Media, Heidelberg, 2012)
7. A. Kahng et al., *VLSI Physical Design: From Graph Partitioning to Timing Closure* (Springer Science & Business Media, Heidelberg, 2011)
8. E. Clarke, J.M. Wing, Formal methods: state of the art and future directions. ACM Comput. Surv. (CSUR) **28**(4), 626–643 (1996)
9. F. Krohm, *The Use of Random Simulation in Formal Verification*. IEEE International Conference on Computer Design (1996)
10. R.E. Bryant, Graph-based algorithms for Boolean function manipulation. IEEE Trans. Comput. **C-35**(8), 677–691 (1986)
11. A. Biere et al. (ed.), *Handbook of Satisfiability*, vol. 185 (IOS Press, Amsterdam, 2009)
12. R.L. Rudell, A. Sangiovanni-Vincentelli, Multiple-valued minimization for PLA optimization. IEEE Trans. CAD **6**(5), 727–750 (1987)
13. R.K. Brayton, G.D. Hachtel, A.L. Sangiovanni-Vincentelli, Multilevel logic synthesis. Proc. IEEE **78**(2), 264–300 (1990)
14. L. Amaru, P.-E. Gaillardon, G. De Micheli, *Majority-Inverter Graph: A Novel Data-Structure and Algorithms for Efficient Logic Optimization*. Design Automation Conference (DAC) (CA, USA, San Francisco, 2014)
15. M. Ciesielski, C. Yu, W. Brown, D. Liu, A. Rossi, *Verification of Gate-level Arithmetic Circuits by Function Extraction*. ACM Design Automation Conference (DAC-2015) (2015)
16. L. Amaru, P.-E. Gaillardon, R. Wille, G. De Micheli, Exploiting Inherent Characteristics of Reversible Circuits for Faster Combinational Equivalence Checking, DATE'16
17. K. Bernstein et al., Device and architecture outlook for beyond CMOS switches. Proc. IEEE **98**(12), 2169–2184 (2010)
18. T. Ernst, Controlling the polarity of silicon nanowire transistors. Science **340**, 1414 (2013)

19. Y.-M Lin et al., High-performance carbon nanotube field-effect transistor with tunable polarities. IEEE Trans. Nanotechnol. **4**(5), 481–489 (2005)
20. H. Yang et al., Graphene barristor, a triode device with a gate-controlled Schottky barrier. Science **336**, 1140 (2012)
21. S.-L. Li et al., Complementary-like graphene logic gates controlled by electrostatic doping. Small **7**(11), 1552–1556 (2011)
22. S. Iba et al., Control of threshold voltage of organic field-effect transistors with double-gate structures. Appl. Phys. Lett. **87**(2), 023509 (2005)
23. D. Lee et al., Combinational logic design using six-terminal NEM relays. IEEE Trans. Comput.-Aided Des. Integr. Circuits Syst. **32**(5), 653–666 (2013)
24. M. Spencer et al., Demonstration of integrated micro-electro-mechanical relay circuits for VLSI applications. IEEE J. Solid-State Circuits **46**(1), 308–320 (2011)
25. T. Toffoli, *Reversible computing, in Automata, Languages and Programming*, ed. by W. de Bakker, J. van Leeuwen (Springer, Heidelberg, 1980), p. 632. (Technical Memo MIT/LCS/TM-151, MIT Lab. for Comput. Sci)
26. T. Schneider et al., Realization of spin-wave logic gates. Appl. Phys. Lett. **92**(2), 022505 (2008)
27. A. Khitun, K.L. Wang, Nano scale computational architectures with spin wave bus. Superlattices Microstruct. **38**(3), 184–200 (2005)
28. A. Khitun et al., Non-volatile magnonic logic circuits engineering. J. Appl. Phys. **110**, 034306 (2011)
29. E. Linn, R. Rosezin, C. Kügeler, R. Waser, Complementary resistive switches for passive nanocrossbar memories. Nat. Mater. **9**, 403 (2010)
30. E. Linn, R. Rosezin, S. Tappertzhofen, U. Böttger, R. Waser, Beyond von Neumann–logic operations in passive crossbar arrays alongside memory operations. Nanotechnology **23**(305205) (2012)
31. S. Miryala et al., Exploiting the Expressive Power of Graphene Reconfigurable Gates via Post-Synthesis Optimization, in *Proceedings of the GLVSLI'15*
32. I. Amlani et al., Digital logic gate using quantum-dot cellular automata. Science **284**(5412), 289–291 (1999)
33. L. Wei et al., Three-input majority logic gate and multiple input logic circuit based on DNA strand displacement. Nano Lett. **13**(6), 2980–2988 (2013)
34. L. Amaru, P.-E. Gaillardon, S. Mitra, G. De Micheli, New logic synthesis as nanotechnology enabler, in *Proceedings of the IEEE* (2015)
35. A. Mishchenko, S. Chatterjee, R.K. Brayton, DAG-aware AIG rewriting a fresh look at combinational logic synthesis, in *Proceedings of the 43rd Annual Design Automation Conference*, pp. 532–535 (2006)
36. A. Mishchenko et al., Delay optimization using SOP balancing, in *Proceedings of the ICCAD* (2011)
37. A. Mishchenko at al., Using simulation and satisfiability to compute flexibilities in Boolean networks. IEEE TCAD **25**(5), 743–755 (2006)
38. O. Coudert, J.C. Madre, A unified framework for the formal verification of sequential circuits. *Proceedings of the ICCAD* (1990)
39. O. Coudert, C. Berthet, J.C. Madre, Verification of sequential machines using Boolean functional vectors, in *Proceedings of the International Workshop on Applied Formal Methods for Correct VLSI Design* (1989)
40. A. Mishchenko et al., Improvements to combinational equivalence checking. *IEEE/ACM International Conference on Computer-Aided Design, ICCAD'06* (2006)

Part I
Logic Representation, Manipulation and Optimization

The first part of this book is dedicated to logic representation, manipulation, and optimization. It deals with two main topics: biconditional logic and majority logic. For biconditional logic, a new canonical binary decision diagram is introduced, examining two variables per decision node rather than only one. For majority logic, a directed-acyclic graph consisting of three-input majority nodes and regular/complemented edges is presented, together with a native Boolean algebra.

Chapter 2
Biconditional Logic

In this chapter, we present *Biconditional Binary Decision Diagrams* (BBDDs), a novel canonical representation form for Boolean functions. BBDDs are binary decision diagrams where the branching condition, and its associated logic expansion, is biconditional on two variables. Empowered by reduction and ordering rules, BBDDs are remarkably compact and unique for a Boolean function. The interest of such representation form in modern *Electronic Design Automation* (EDA) is twofold. On the one hand, BBDDs improve the efficiency of traditional EDA tasks based on decision diagrams, especially for arithmetic intensive designs. On the other hand, BBDDs represent the natural and native design abstraction for emerging technologies where the circuit primitive is a comparator, rather than a simple switch. We provide, in this chapter, a solid ground for BBDDs by studying their underlying theory and manipulation properties. Thanks to an efficient BBDD software package implementation, we validate (i) runtime reduction in traditional decision diagrams applications with respect to other DDs, and (ii) improved synthesis of circuits in standard and emerging technologies.

2.1 Introduction

The choice of data structure is crucial in computing applications, especially for the automated design of digital circuits. When logic functions are concerned, *Binary Decision Diagrams* (BDDs) [1–3] are a well established cogent and unique, i.e., canonical, logic representation form. BDDs are widely used in *Electronic Design Automation* (EDA) to accomplish important tasks, e.g., synthesis [4], verification [5], testing [6], simulation [7], and others. Valuable extensions [8] and generalizations [9] of BDDs have been proposed in literature to improve the performance of EDA applications based on decision diagrams. The corresponding software packages [10, 11] are indeed mature and supported by a solid theory. However, there are still combinational designs, such as multipliers and arithmetic circuits, that do not fit modern computational capabilities when represented by existing canonical

© Springer International Publishing Switzerland 2017
L.G. Amaru, *New Data Structures and Algorithms for Logic Synthesis and Verification*, DOI 10.1007/978-3-319-43174-1_2

decision diagrams [24]. The quest for new data structures handling such hard circuits, and possibly pushing further the performance for ordinary circuits, is of paramount importance for next-generation digital designs. Furthermore, the rise of emerging technologies carrying new logic primitives demands for novel logic representation forms that fully exploit a diverse logic expressive power. For instance, controllable polarity *Double-Gate* (DG) transistors, fabricated in silicon nanowires [12], carbon nanotubes [13] or graphene [14] technologies, but also nanorelays [15], intrinsically behave as comparators rather than switches. Hence, conventional data structures are not appropriate to model natively their functionality [16].

In this chapter, we present *Biconditional Binary Decision Diagrams* (BBDDs), a novel canonical BDD extension. While original BDDs are based on the single-variable Shannon's expansion, BBDDs employ a two-variable biconditional expansion, making the branching condition at each decision node dependent on two variables per time. Such feature improves the logic expressive power of the binary decision diagram. Moreover, BBDDs represent also the natural and native design abstraction for emerging technologies [12–15] where the circuit primitive is a comparator, rather than a switch.

We validate the benefits deriving from the use of BBDDs in EDA tasks through an efficient software manipulation package, available online [19]. Considering the MCNC benchmark suite, BBDDs are built $1.4\times$ and $1.5\times$ faster than original BDDs and *Kronecker Functional Decision Diagrams* (KFDDs) [9], while having also $1.5\times$ and $1.1\times$ fewer nodes, respectively. Moreover, we show hard arithmetic circuits that fit computing capabilities with BBDDs but are not practical with state-of-art BDDs or KFDDs. Employed in the synthesis of an iterative decoder design, targeting standard CMOS technology, BBDDs advantageously pre-structure arithmetic circuits as front-end to a commercial synthesis tool, enabling to meet tight timing constraints otherwise beyond the capabilities of traditional synthesis. The combinational verification of the optimized design is also sped up by 11.3% using BBDDs in place of standard BDDs. Regarding the automated design for emerging technologies, we similarly employed BBDDs as front-end to a commercial synthesis tool but then targeting a controllable-polarity *Double-Gate* (DG) *Silicon NanoWires Field Effect Transistors* (SiNWFETs) technology [12]. Controllable-polarity DG-SiNWFETs behave as binary comparators [12]. Such primitive is naturally modelled by BBDDs. Experimental results show that the effectiveness of BBDD pre-structuring for circuits based on such devices is even higher than for standard CMOS, thus enabling a superior exploitation of the emerging technology features.

The remainder of this chapter is organized as follows. Section 2.2 first provides a background on BDDs and then discusses the motivations for the study of BBDDs. In Sect. 2.3, the formal theory for BBDDs is introduced, together with efficient manipulation algorithms. Section 2.4 first shows theoretical size bounds for notable functions represented with BBDDs and then compares the performance of our BBDD software package with other state-of-art packages for BDDs and KFDDs. Section 2.5 presents the application of BBDDs to circuit synthesis and verification in traditional technology. Section 2.6 presents the application of BBDDs to circuit synthesis in emerging technologies. This chapter is concluded in Sect. 2.7.

2.2 Background and Motivation

This section first provides the background and the basic terminology associated with
binary decision diagrams and their extensions. Then, it discusses the motivations to
study BBDDs, from both a traditional EDA and an emerging technology perspectives.

2.2.1 Binary Decision Diagrams

Binary Decision Diagrams (BDDs) are logic representation structures first introduced by Lee [1] and Akers [2]. Ordering and reduction techniques for BDDs were
introduced by Bryant in [3] where it was shown that, with these restrictions, BDDs
are a canonical representation form. Canonical BDDs are often compact and easy to
manipulate. For this reason, they are extensively used in EDA and computer science.
In the following, we assume that the reader is familiar with basic concepts of Boolean
algebra (for a review see [1, 21]) and we review hereafter the basic terminology used
in the rest of the paper.

2.2.1.1 Terminology and Fundamentals

A BDD is a *Direct Acyclic Graph* (DAG) representing a Boolean function. A BDD
is uniquely identified by its *root*, the set of *internal nodes*, the set of *edges* and the
1/0-sink terminal nodes. Each internal node (Fig. 2.1a) in a BDD is labeled by a
Boolean variable v and has two out-edges labeled 0 and 1.

Fig. 2.1 BDD *non-terminal*
node (**a**) canonical BDD for
$a \cdot b$ function (**b**)

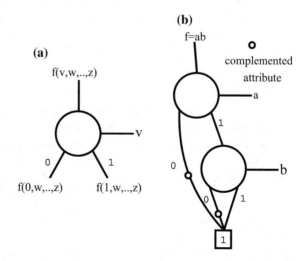

Each internal node also represents the Shannon's expansion with respect to its variable v:

$$f(v, w, \ldots, z) = v \cdot f(1, w, \ldots, z) + v' \cdot f(0, w, \ldots, z) \qquad (2.1)$$

The 1- and 0-edges connect to positive and negative Shannon's cofactors, respectively.

Edges are characterized by a regular/complemented attribute. Complemented edges indicate to invert the function pointed by that edge.

We refer hereafter to BDDs as to *canonical reduced and ordered* BDDs [3], that are BDDs where (i) each input variable is encountered at most once in each root to sink path and in the same order on all such paths, (ii) each internal node represent a distinct logic function and (iii) only 0-edges can be complemented. Figure 2.1b shows the BDD for function $f = a \cdot b$.

In the rest of this paper, symbols $\oplus$ and $\odot$ represent XOR and XNOR operators, respectively. Symbol $\otimes$ represents any 2-operand Boolean operator.

2.2.1.2 Previous BDD Extensions

Despite BDDs are typically very compact, there are functions for which their representation is too large to be stored and manipulated. For example, it was shown in [24] that the BDD for the multiplier of two n-bit numbers has at least $2^{n/8}$ nodes. For this reason, several extensions of BDDs have been suggested.

One first extension are free BDDs, where the variable order condition is relaxed allowing polynomial size representation for the multiplier [22]. However, such relaxation of the order sacrifices the canonicity of BDDs, making manipulation of such structures less efficient. Indeed, canonicity is a desirable property that permits operations on BDDs to have an efficient runtime complexity [3]. Another notable approach trading canonicity for compactness is parity-BDDs ($\oplus$-BDDs) presented in [25]. In $\oplus$-BDDs, a node can implement either the standard Shannon's expansion or the $\oplus$ (XOR) operator. Thanks to this increased flexibility, $\oplus$-BDDs allow certain functions having exponential size with original BDDs to be instead represented in polynomial size. Again, the manipulation of $\oplus$-BDDs is not as efficient as with original BDDs due to the heterogeneity introduced in the diagrams by additional $\oplus$-nodes.

Considering now BDD extensions preserving canonicity, zero-suppressed BDDs [30] are BDDs with modified reduction rules (node elimination) targeting efficient manipulation of sparse sets. *Transformation BDDs* (TBDDs) [33, 35] are BDDs where the input variables of the decision diagram are determined by a logic transformation of the original inputs. When the input transformation is an injective mapping, TBDDs are canonical representation form [35]. In theory, TBDDs can represent every logic function with polynomial size given the perfect input transformation. However, the search for the perfect input transformation is an intractable problem. Moreover, traditional decision diagram manipulation algorithms (e.g., variable re-ordering) are not efficient with general TBDDs due to the presence of the input transformation

[22]. Nevertheless, helpful and practical TBDDs have been proposed in literature, such as linear sifting of BDDs [31, 32] and *Hybrid Decision Diagrams* (HDDs) [34]. Linear sifting consists of linear transformations between input variables carried out on-line during construction. The linear transformations are kept if they reduce the size of the BDD or undone in the other case. On the one hand, this makes the linear transform dependent itself on the considered BDD and therefore very effective to reduce its size. On the other hand, different BDDs may have different transforms and logic operations between them become more complicated. More discussion for linear sifting and comparisons to our proposed BDD extension are given in Sect. 2.3.1. HDDs are TBDDs having as transformation matrix the Kronecker product of different 2×2 matrices. The entries of such matrices are determined via heuristic algorithms. HDDs are reported to achieve a remarkable size compression factor (up to 3 orders of magnitude) with respect to BDDs [34] but they suffer similar limitations as linear sifting deriving from the dependency on the particular, case-dependent, input transformation employed.

Other canonical extensions of BDDs are based on different core logic expansions driving the decision diagram. *Functional Decision Diagrams* (FDDs) [8] fall in this category employing the (positive) Davio's expansion in place of the Shannon's one:

$$f(v, w, \ldots, z) = f(0, w, \ldots, z) \oplus v \cdot (f(0, w, \ldots, z) \oplus f(1, w, \ldots, z)) \quad (2.2)$$

Since the Davio expansion is based on the $\oplus$ operator, FDDs provide competitive representations for XOR-intensive functions. *Kronecker FDDs* (KFDDs) [9] are a canonical evolution of FDDs that can employ both Davio's expansions (positive and negative) and Shannon's expansion in the same decision diagram provided that all the nodes belonging to the same level use the same decomposition type. As a consequence, KFDDs are a superset of both FDDs and BDDs. However, the heterogeneity of logic expansion types employable in KFDDs makes their manipulation slightly more complicated than with standard BDDs. For problems that are more naturally stated in the discrete domain rather than in terms of binary values, *Multi-valued Decision Diagrams* (MDDs) have been proposed [40] as direct extension of BDDs. MDDs have multiple edges, as many as the cardinality of the function domain, and multiple sink nodes, as many as the cardinality of the function codomain. We refer the reader to [22] for more details about MDDs.

Note that the list of BDD extensions considered above is not complete. Due to the large number of extensions proposed in literature, we have discussed only those relevant for the comprehension of this work.

In this chapter, we present a novel canonical BDD extension where the branching decisions are biconditional on two variables per time rather than on only one. The motivation for this study is twofold. First, from a theoretical perspective, considering two variables per time enhances the expressive power of a decision diagram. Second, from an application perspective, there exist emerging devices better modeled by a two-variable (biconditional) comparator rather than a single variable switch. In this context, the proposed BDD extension serves as natural logic abstraction. A discussion about the technology motivation for this work is provided hereafter.

2.2.2   Emerging Technologies

Many logic representation forms are inspired by the underlying functionality of contemporary digital circuits. Silicon-based *Metal-Oxide-Semiconductor Field-Effect Transistors* (MOSFETs) form the elementary blocks for present electronics. In the digital domain, a silicon transistor behaves as a two-terminal binary switch driven by a single input signal. The Shannon's expansion captures such operation in the form of a Boolean expression. Based on it, logic representation and manipulation of digital circuits is efficient and automated.

With the aim to support the exponential growth of digital electronics in the future, novel elementary blocks are currently under investigation to overcome the physical limitations of standard transistors. Deriving from materials, geometries and physical phenomena different than MOSFETs, many emerging devices are not naturally modeled by traditional logic representation forms. Therefore, novel CAD methodologies are needed, which appropriately handle such emerging devices.

We concentrate here on a promising class of emerging devices that inherently implement a two-input comparator rather than a simple switch. These innovative devices come in different technologies, such as silicon nanowires [12], carbon nanotubes [13], graphene [14] and nanorelays [15]. In the first three approaches, the basic element is a double-gate controllable-polarity transistor. It enables online configuration of the device polarity (*n* or *p*) by adjusting the voltage at the second gate. Consequently, in such a double-gate transistor, the *on/off* state is biconditional on both gates values. The basic element in the last approach [15] is instead a six-terminals nanorelays. It can implement complex switching functions by controlling the voltages at the different terminals. Following to its geometry and physics, the final electric way connection in the nanorelay is biconditional on the terminal values [15]. Even though they are based on different technologies, all the devices in [12–15] have the same common logic abstraction, depicted by Fig. 2.2.

In this chapter, we mainly focus on double-gate controllable polarity SiNWFETs [12] to showcase the impact of novel logic representation forms in emerging technology synthesis. A device sketch and fabrication views from [12] are reported in Fig. 2.3 for the sake of clarity.

Fig. 2.2 Common logic abstraction for emerging devices: controllable polarity double-gate FETs in silicon nanowires [12], carbon nanotubes [13], graphene [14] but also six terminal nanorelays [15]

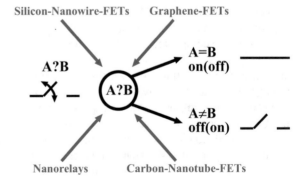

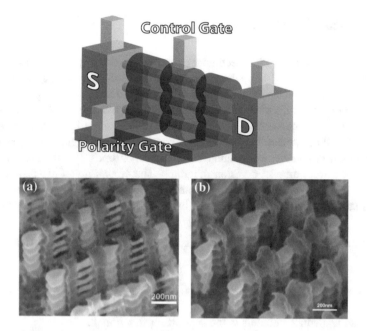

Fig. 2.3 Sketch structure and fabrication images of controllable polarity double-gate SiNWFETs from [12]

Then, we also present results for two other nanotechnologies featuring a two-input comparator functionality: nanorelays [15] and reversible logic [55].

Without a dedicated logic abstraction and synthesis methodology, the full potential of these technologies may remain unveiled. We propose in this paper a novel logic representation form, based on the biconditional connective, that naturally harnesses the operation of a two-input comparator. Section 2.6 will show the impact of our representation form in the synthesis of emerging nanotechnologies.

2.3 Biconditional Binary Decision Diagrams

This section introduces *Biconditional Binary Decision Diagrams* (BBDDs). First, it presents the core logic expansion that drives BBDDs. Then, it gives ordering and reduction rules that makes *Reduced and Ordered BBDDs* (ROBBDDs) compact and canonical. Finally, it discusses efficient algorithms for BBDD manipulation and their practical implementation in a software package.

2.3.1 Biconditional Expansion

Logic expansions, also called decompositions, are the driving core of various types of decision diagrams. In [42], a theoretical study concluded that, among all the possible one-variable expansions, only Shannon's, positive Davio's and negative Davio's types help to reduce the size of decision diagrams. While this result prevents from introducing more one-variable decomposition types, new multi-variable decompositions are still of interest. In this work, we consider a novel logic expansion, called *biconditional expansion*, examining two variables per time rather than one, in order to produce novel compact decision diagrams. The *biconditional expansion* is one of the many possible two-variable decompositions. Note that other advantageous two-variable decompositions may exist but their study is out of the scope of this work.

Definition 2.1 The *biconditional expansion* is a two-variable expansion defined $\forall f \in \mathbb{B}^n$, with $n > 1$, as:

$$f(v, w, \ldots, z) = (v \oplus w) \cdot f(w', w, \ldots, z) + (v \odot w) \cdot f(w, w, \ldots, z) \qquad (2.3)$$

with v and w distinct elements in the support for function f.

As per the *biconditional expansion* in (2.3), only functions that have two or more variables can be decomposed. Indeed, in single variable functions, the terms $(v \oplus w)$ and $(v \odot w)$ cannot be computed. In such a condition, the *biconditional expansion* of a single variable function can reduce to a Shannon's expansion by fixing the second variable w to logic 1. With this boundary condition, any Boolean function can be fully decomposed by *biconditional expansions*.

Note that a similar concept to *biconditional expansion* appears in [31, 32] where linear transformations are applied to BDDs. The proposed transformation replaces one variable x_i with $x_i \odot x_j$. In the BDD domain, $x_i \mapsto x_i \odot x_j$ transforms a Shannon's expansion around variable x_i into a *biconditional expansion* around variables x_i and x_j. We differentiate our work from linear transformations by the abstraction level at which we embed the biconditional connective. Linear transformations in [31, 32] operate as post-processing of a regular BDD, while we propose to entirely substitute the Shannon's expansion with the *biconditional expansion*. By changing the core engine driving the decision diagram new compact representation opportunities arise. However, a solid theoretical foundation is needed to exploit such potential. We address this requirement in the rest of this section.

2.3.2 BBDD Structure and Ordering

Biconditional Binary Decision Diagrams (BBDD) are driven by the *biconditional expansion*. Each non-terminal node in a BBDD has the branching condition *biconditional* on two variables. We call these two variables the *Primary Variable* (PV) and

Fig. 2.4 BBDD
non-terminal node

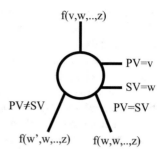

the *Secondary Variable* (SV). An example of a BBDD *non-terminal* node is provided by Fig. 2.4. We refer hereafter to $PV \neq SV$ and $PV = SV$ edges in a BBDD node simply as $\neq$-*edges* and =-*edges*, respectively.

To achieve *Ordered BBDDs* (OBBDDs), a variable order must be imposed for PVs and a rule for the other variables assignment must be provided. We propose the *Chain Variable Order* (CVO) to address this task. Given a Boolean function f and a variable order $\pi = (\pi_0, \pi_1, \ldots, \pi_{n-1})$ for the support variables of f, the CVO assigns PVs and SVs as:

$$\begin{cases} PV_i = \pi_i \\ SV_i = \pi_{i+1} \end{cases} \text{with } i = 0, 1, \ldots, n-2; \begin{cases} PV_{n-1} = \pi_{n-1} \\ SV_{n-1} = 1 \end{cases} \quad (2.4)$$

Example 2.1 CVO Example: From $\pi = (\pi_0, \pi_1, \pi_2)$, the corresponding CVO ordering is obtained by the following method. First, $PV_0 = \pi_0$, $PV_1 = \pi_1$ and $SV_0 = \pi_1$, $SV_1 = \pi_2$ are assigned. Then, the final boundary conditions of (2.4) are applied as $PV_2 = \pi_2$ and $SV_2 = 1$. The consecutive ordering by couples (PV_i, SV_i) is thus $((\pi_0, \pi_1), (\pi_1, \pi_2), (\pi_2, 1))$.

The *Chain Variable Order* (CVO) is a key factor enabling unique representation of ordered biconditional decision diagrams. For the sake of clarity, we first consider the effect of the CVO on *complete* OBBDDs and then we move to generic reduced BBDDs in the next subsection.

Definition 2.2 A *complete* OBBDD of n variables has 2^n-1 distinct internal nodes, no sharing, and 2^n terminal 0–1 nodes.

Lemma 2.1 *For a Boolean function f and a variable order π, there exists only one complete OBBDD ordered with CVO(π).*

Proof Say n the number of variables in f. All *complete* OBBDD of n variables have an identical internal structure, i.e., a full binary tree having $2^n - 1$ internal nodes. The distinctive feature of a *complete* OBBDD for f is the distribution of terminal 0–1 nodes. We need to show that such distribution is unique in a *complete* OBBDD ordered with CVO(π). Consider the unique truth table for f with 2^n elements filled as per π. Note that in a *complete* OBBDD there are 2^n distinct paths by construction.

We link the terminal value reached by each path to an element of the truth table. We do so by recovering the binary assignment of π generating a path. That binary assignment is the linking address to the truth table entry. For example, the terminal value reached by the path $(\pi_0 \neq \pi_1, \pi_1 = \pi_2, \pi_2 \neq 1)$ corresponds to the truth table entry at the address $(\pi_0 = 1, \pi_1 = 0, \pi_2 = 0)$. Note that distinct paths in the CVO(π) corresponds to distinct binary assignments of π, owing to the isomorphism induced by the *biconditional expansion*. By exhausting all the 2^n paths we are guaranteed to link all entries in the truth table. This procedure establishes a one-to-one correspondence between the truth table and the *complete* OBBDD. Since truth tables filled as per π are unique, also *complete* OBBDD ordered with CVO(π) are unique.   ∎

We refer hereafter to OBBDDs as to BBDDs ordered by the CVO.

2.3.3 BBDD Reduction

In order to improve the representation efficiency, OBBDDs should be reduced according to a set of rules. We present hereafter BBDD reduction rules, and we discuss the uniqueness of the so obtained ordered and reduced BBDDs.

2.3.3.1 Reduction Rules

The straightforward extension of OBDD reduction rules [3] to OBBDDs, leads to *weak reduced OBBDDs* (ROBBDDs). This kind of reduction is called *weak* due to the partial exploitation of OBBDD reduction opportunities. A *weak* ROBBDD is an OBBDD respecting the two following rules:
(**R1**) It contains no two nodes, root of isomorphic subgraphs.
(**R2**) It contains no nodes with identical children.
   In addition, the OBBDD representation exhibits supplementary interesting features enabling further reduction opportunities. First, levels with no nodes (empty levels) may occur in OBBDDs. An empty level is a level in the decision diagram created by the Chain Variable Order but containing no nodes as a result of the augmented functionality of the *biconditional expansion*. Such levels must be removed to compact the original OBBDD. Second, subgraphs that represent functions of a single variable degenerates into a single DD node driven by the Shannon's expansion followed by the sink terminal node. The degenerated node functionality is the same as in a traditional BDD node. Single variable condition is detectable by checking the cardinality of the support set of the subgraph.
   Formally, a *strong* ROBBDD is an OBBDD respecting **R1** and **R2** rules, and in addition:
(**R3**) It contains no empty levels.
(**R4**) Subgraphs representing single variable functions degenerates into a single DD node driven by the Shannon's expansion.

Fig. 2.5 Function to be
represented:
$f = a \cdot b + (a \oplus b) \cdot (c \odot d)$,
weak ROBBDD for f (**a**)
and *strong* ROBBDD
for f (**b**)

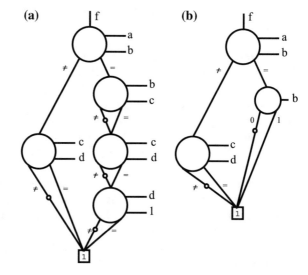

For the sake of simplicity, we refer hereafter to a single variable subgraph degenerated into a single DD node as a BDD node.

Figure 2.5 depicts *weak* and *strong* ROBBDDs for the function $f = a \cdot b + (a \oplus b) \cdot (c \odot d)$. The *weak* ROBBDD is forced to allocate 4 levels (one for each variable) to fully represent the target function resulting in 5 internal nodes. On the other hand, the *strong* ROBBDD exploits reduction rule **R4** collapsing the =-branch of the root node ($a = b$) into a single BDD node. Moreover, rule **R3** suppresses empty level further compressing the diagram in 3 levels of depth and 3 internal nodes.

2.3.3.2 Canonicity

Weak and *strong* reduced OBBDDs are canonical, as per:

Lemma 2.2 *For a given Boolean function f and a variable order π, there exists only one* weak *ROBBDD.*

Proof Weak ROBBDDs are obtained by applying reduction rules **R1** and **R2**, in any combination, to an OBBDD until no other **R1** or **R2** rule can be applied. Without loss of generality, suppose to start from a *complete* OBBDD. Any other valid OBBDD can be reached during the reduction procedure. In [44], it is shown that the iterative reduction of general decision diagrams, based on rules **R1** and **R2**, reaches a unique structure. Since the initial *complete* OBBDD is unique, owing to Lemma 2.1, and the iterative reduction based on rules **R1** and **R2** leads to a unique outcome, owing to [44], also *weak* ROBBDD are unique for a CVO(π), i.e., canonical. ■

Theorem 2.1 *A strong ROBBDD is a canonical representation for any Boolean function f.*

Proof strong ROBBDDs can be directly derived by applying reduction rules **R3** and **R4**, in any combination, to *weak* ROBBDDs until no other **R3** or **R4** rule can be applied.

In order to prove the canonicity of *strong* ROBBDD, we proceed by five succeeding logical steps. The final goal is to show that any sequence of reductions drawn from {**R3,R4**}, that continues until no other reduction is possible, reaches a unique *strong* ROBBDD structure, preserving the uniqueness property of the starting *weak* ROBBDD.

1. Reductions **R3** and **R4** preserve distinctness. As it holds for rules **R1** and **R2**, also **R3** and **R4** preserve distinctness. Rule **R3** compacts the decision diagram structure without any ambiguity in the elimination of levels, i.e., when a level is empty it is uniquely removed. Rule **R4** substitutes single variable functions with a single BDD node (followed by the sink node). This operation has a specific and unique outcome since it is combined with rules **R1** and **R2** (each node represents a distinct logic function).

2. The set of applicable rules **R4** is fixed. In a given *weak* ROBBDD, the set of all possible single variable subgraph collapsing (rule **R4**) is fixed *a priori*, i.e., there exists a specific set of applicable **R4** reductions independent of the reduction sequence employed. Consider a top-down exploration of the starting *weak* ROBBDD. At each branching condition, the support sets of the current node children are checked. If the cardinality of the support set is 1 (single variable) then this subgraph is reducible by **R4**. Regardless of the particular exploration order, the support set of all subgraphs remains the same. Therefore, the applicability of rules **R4** depends only on the given *weak* ROBBDD structure.

3. Rules **R4** are independent of rules **R3**. Rules **R3** (empty levels elimination) cannot preclude the exercise of rules **R4** (single-variable subgraphs collapsing) because they eliminate levels with no nodes, where no rule **R4** could apply.

4. Rules **R3** can be precluded by rules **R4**. Rules **R4** can preclude the exercise of rules **R3** since the collapse of subgraphs into a single node can make some levels in the decision diagram empty (see Fig. 2.5). Nevertheless, each rule **R3** is reachable in a reduction sequence that guarantees to exhaust all the blocking **R4** before its termination.

5. Iterative reduction strategy is order independent. We refer to an iterative reduction strategy as to a sequence of reductions drawn from {**R3**, **R4**} applied to a *weak* ROBBDD, that continues until no other reduction is possible. At each step of reduction sequence, the existence of a new reduction **R3** or **R4** is checked. Following points 2 and 3, all possible **R4** are identifiable and reachable at any time before the end of the reduction sequence, regardless of the order employed. Consider now rules **R3**. Some of them are not precluded by rules **R4**. Those are also identifiable and reachable at any time before the end of the reduction sequence. The remaining **R3** are precluded by some **R4**. However, all possible **R4**, included those blocking some **R3**, are guaranteed to be accomplished before the end of the reduction. Therefore, there

always exists a step, in any reduction sequence, when each rule **R3** is reachable as the blocking **R4** are exhausted. Consequently, any iterative reduction strategy drawn from {**R3**, **R4**} achieves a unique reduced BBDD structure (*strong* ROBBDD).

It follows that any combination of reduction rules **R3** and **R4** compact a canonical *weak* ROBBDD into a unique *strong* ROBBDD, preserving canonicity. ∎

2.3.4 BBDD Complemented Edges

Being advantageously applied in modern ROBDD2s packages [10], complemented edges indicate to invert the function pointed by an edge. The canonicity is preserved when the complement attribute is allowed only at 0-edges (only logic 1 terminal node available). Reduction rules **R1** and **R2** continue to be valid with complemented edges [22]. Similarly, we extend ROBBDDs to use complemented edges only at $\neq$-edges, with also only logic 1 terminal node available, to maintain canonicity.

Theorem 2.2 *ROBBDDs with complemented edges allowed only at $\neq$-edges are canonical.*

Proof Reduction rules **R1** and **R2** support complemented edges at the else branch of canonical decision diagrams [22]. In BBDDs, the else branch is naturally the $\neq$-edge, as the *biconditional* connective is true (then branch) with the =-edge. We can therefore extend the proof of Lemma 2.2 to use complemented edges at $\neq$-edges and to remove the logic 0 terminal node. It follows that *weak* ROBBDDs with complented edges at $\neq$-edges are canonical. The incremental reduction to *strong* ROBBDDs does not require any knowledge or action about edges. Indeed, the proof of Theorem 2.1 maintains its validity with complemented edges. Consequently, *strong* ROBBDDs with complemented edges at $\neq$-edges are canonical. ∎

For the sake of simplicity, we refer hereafter to BBDDs as to canonical ROBBDDs with complemented edges, unless specified otherwise.

2.3.5 BBDD Manipulation

So far, we showed that, under ordering and reduction rules, BBDDs are unique and potentially very compact. In order to exploit such features in real-life tools, a practical theory for the construction and manipulation of BBDDs is needed. We address this requirement by presenting an efficient manipulation theory for BBDDs with a practical software implementation, available online at [19].

2.3.5.1 Rationale for Construction and Manipulation of BBDDs

DDs are usually built starting from a netlist of Boolean operations. A common strategy employed for the construction task is to build bottom-up the DD for each element in the netlist, as a result of logic operations between DDs computed in the previous steps. In this context, the core of the construction task is an efficient Boolean operation algorithm between DDs. In order to make such approach effective in practice, other tasks are also critical, such as memory organization and re-ordering of variables. With BBDDs, we follow the same construction and manipulation rationale, but with specialized techniques taking care of the *biconditional expansion*.

2.3.5.2 Considerations to Design an Efficient BBDD Package

Nowadays, one fundamental reason to keep decision diagrams small is not just to successfully fit them into the memory, that in a modern server could store up to 1 billion nodes, but more to maximize their manipulation performance. Following this trend, we design the BBDD manipulation algorithms and data structures aiming to minimize the runtime while keeping under control the memory footprint. The key concepts unlocking such target are (i) *unique* table to store BBDD nodes in a *strong canonical form*,[1] (ii) recursive formulation of Boolean operations in terms of *biconditional expansions* with relative *computed* table, (iii) memory management to speed up computation and (iv) *chain* variable re-ordering to minimize the BBDD size. We discuss in details each point hereafter.

2.3.5.3 Unique Table

BBDD nodes must be stored in an efficient form, allowing fast lookup and insertion. Thanks to canonicity, BBDD nodes are uniquely labeled by a tuple {CVO-level, ≠-child, ≠-attribute, =-child}. A *unique* table maps each tuple {CVO-level, ≠-child, ≠-attribute, =-child} to its corresponding BBDD node via a hash-function. Hence, each BBDD node has a distinct entry in the *unique* table pointed by its hash-function, enabling a *strong canonical form* representation for BBDDs.

Exploiting this feature, equivalence test between two BBDD nodes corresponds to a simple pointer comparison. Thus, lookup and insertion operations in the *unique* table are efficient. Before a new node is added to the BBDD, a lookup checks if its corresponding tuple {CVO-level, ≠-child, ≠-attribute, =-child} already exists in the *unique* table and, if so, its pointed node is returned. Otherwise, a new entry for the node is created in the *unique* table.

---

[1] A strong canonical form is a form of data pre-conditioning to reduce the complexity of equivalence test [46].

2.3.5.4 Boolean Operations Between BBDDs

The capability to apply Boolean operations between two BBDDs is essential to represent and manipulate large combinatorial designs. Consequently, an efficient algorithm to compute $f \otimes g$, where $\otimes$ is any Boolean function of two operands and $\{f, g\}$ are two existing BBDDs, is the core of our manipulation package. A recursive formulation of $f \otimes g$, in terms of *biconditional* expansions, allows us to take advantage of the information stored in the existing BBDDs and hence reduce the computation complexity of the successive operation. Algorithm 1 shows the outline of the recursive implementation for $f \otimes g$. The input of the algorithm are the BBDDs for $\{f, g\}$, and the two-operand Boolean function $\otimes$ that has to be computed between them. If f and g are identical, or one of them is the sink 1 node, the operation $f \otimes g$ reaches a terminal condition. In this case, the result is retrieved from a pre-defined list of trivial operations and returned immediately (Alg. 1α). When a terminal condition is not encountered, the presence of $\{f, g, \otimes\}$ is first checked in a *computed* table, where previously performed operations are stored in case of later use. In the case of positive outcome, the result is retrieved from the *computed* table and returned immediately (Alg. 1β). Otherwise, $f \otimes g$ has to be explicitly computed (Alg. 1γ). The top level in the CVO for $f \otimes g$ is determined as $i = max_{level}\{f, g\}$ with its $\{PV_i, SV_i\}$ referred as to $\{v, w\}$, respectively, for the sake of simplicity. The root node for $f \otimes g$ is placed at such level i and its children computed recursively. Before proceeding in this way, we need to ensure that the two-variable *biconditional* expansion is well defined for both f and g, particularly if they are single variable functions. To address this case, single variable functions are prolonged down to $min_{level}\{f, g\}$ through a chain of consecutive BBDD nodes. This temporarily, and locally, may violate reduction rule **R4** to guarantee consistent $\neq$- and =-edges. However, rule **R4** is enforced before the end of the algorithm. Provided such handling strategy, the following recursive formulation, in terms of *biconditional* expansions, is key to efficiently compute the children for $f \otimes g$:

$$f \otimes g = (v \oplus w)(f_{v \neq w} \otimes g_{v \neq w}) + (v \overline{\oplus} w)(f_{v=w} \otimes g_{v=w}) \qquad (2.5)$$

The term $(f_{v \neq w} \otimes g_{v \neq w})$ represents the $\neq$-child for the root of $f \otimes g$ while the term $(f_{v=w} \otimes g_{v=w})$ represents the =-child. In $(f_{v \neq w} \otimes g_{v \neq w})$, the Boolean operation $\otimes$ needs to be updated according to the regular/complemented attributes appearing in the edges connecting to $f_{v \neq w}$ and $g_{v \neq w}$. After the recursive calls for $(f_{v=w} \otimes g_{v=w})$ and $(f_{v \neq w} \otimes g_{v \neq w})$ return their results, reduction rule **R4** is applied. Finally, the tuple $\{top\text{-}level, \neq\text{-}child, \neq\text{-}attribute, =\text{-}child\}$ is found or added in the *unique* table and its result updated in the *computed* table.

Algorithm 1 : $f \otimes g$

INPUT: BBDDs for $\{f, g\}$ and Boolean operation $\otimes$.
OUTPUT: BBDD top node R for $f \otimes g$, edge attribute $(Attr)$ for $f \otimes g$.

if (terminal case)$||(f == g)$ **then** } α
  $\{R, Attr\}$ = identical_terminal($\{f, g, \otimes\}$);
  return $\{R, Attr\}$;
else if *computed* table has entry $\{f, g, \otimes\}$ **then** } β
  $\{R, Attr\}$ = lookup *computed* table($\{f, g, \otimes\}$);
  return $\{R, Attr\}$;
else } γ
  $i = max_{level}\{f, g\}$;
  $\{v, w\} = \{PV, SV\}@(level = i)$;
  if $(|supp(f)| == 1)||(|supp(g)| == 1)$ **then**
    chain-transform(f, g);
  end if
  $\{E, E \rightarrow Attr\} = f_{v=w} \otimes g_{v=w}$;
  $\otimes_D = update_{op}(\otimes, f_{v \neq w} \rightarrow Attr, g_{v \neq w} \rightarrow Attr)$;
  $\{D, D \rightarrow Attr\} = f_{v \neq w} \otimes_D g_{v \neq w}$;
  if reduction rule **R4** applies **then**
    R =BDD-node $@(level = i)$;
  else if $\{E, E \rightarrow Attr\} == \{D, D \rightarrow Attr\}$ **then**
    $R = E$;
  else
    $D \rightarrow Attr = update_{attr}(E \rightarrow Attr, D \rightarrow Attr)$;
    $R = lookup_insert(i, D, D \rightarrow Attr, E)$;
  end if
  insert *computed* table $(\{f, g, \otimes\}, R, E \rightarrow Attr)$;
  return $\{R, E \rightarrow Attr\}$;
end if

Observe that the maximum number of recursions in Eq. 2.5 is determined by all possible combination of nodes between the BBDDs for f and g. Assuming constant time lookup in the *unique* and *computed* tables, it follows that the time complexity for Algorithm 1 is $O(|f| \cdot |g|)$, where $|f|$ and $|g|$ are the number of nodes of the BBDDs of f and g, respectively.

2.3.5.5 Memory Management

The software implementation of data-structures for *unique* and *computed* tables is essential to control the memory footprint but also to speed-up computation. In traditional logic manipulation packages [10], the *unique* and *computed* tables are implemented by a hash-table and a cache, respectively. We follow this approach in the BBDD package, but we add some specific additional technique. Informally, we minimize the access time to stored nodes and operations by dynamically changing the data-structure size and hashing function, on the basis of a $\{$size$\times$access-time$\}$ quality metric.

The core hashing-function for all BBDD tables is the Cantor pairing function between two integer numbers [45]:

$$C(i, j) = 0.5 \cdot (i + j) \cdot (i + j + 1) + i \qquad (2.6)$$

which is a bijection from $\mathbb{N}_0 \times \mathbb{N}_0$ to $\mathbb{N}_0$ and hence a *perfect hashing function* [45]. In order to fit the memory capacity of computers, modulo operations are applied after the Cantor pairing function allowing collisions to occur. To limit the frequency of collisions, a first modulo operation is performed with a large prime number m, e.g., $m = 15485863$, for statistical reasons. Then, a final modulo operation resizes the result to the current size of the table.

Hashing functions for *unique* and *computed* tables are obtaining by nested Cantor pairings between the tuple elements with successive modulo operations.

Collisions are handled in the *unique* table by a linked list for each hash-value, while, in the *computed* table, the cache-like approach overwrites an entry when collision occurs.

Keeping low the frequency of collisions in the *unique* and *computed* tables is of paramount importance to the BBDD package performance. Traditional garbage collection and dynamic table resizing [10] are used to address this task. When the benefit deriving by these techniques is limited or not satisfactory, the hash-function is automatically modified to re-arrange the elements in the table. Standard modifications of the hash-function consist of nested Cantor pairings re-ordering and re-sizing of the prime number m.

2.3.5.6 Chain Variable Re-Ordering

The chain variable order for a BBDD influences the representation size and therefore its manipulation complexity. Automated chain variable re-ordering assists the BBDD package to boost the performance and reduce the memory requirements. Efficient reordering techniques for BDDs are based on local variable swap [47] iterated over the whole variable order, following some minimization goal. The same approach is also efficient with BBDDs. Before discussing on convenient methods to employ local swaps in a global reordering procedure, we present a new core variable swap operation adapted to the CVO of BBDDs.

BBDD CVO Swap: Variable swap in the CVO exchanges the PVs of two adjacent levels i and $i+1$ and updates the neighbor SVs accordingly. The effect on the original variable order π, from which the CVO is derived as per Eq. 2.4, is a direct swap of variables π_i and π_{i+1}. Note that all the nodes/functions concerned during a CVO swap are overwritten (hence maintaining the same pointer) with the new tuple generated at the end of the operation. In this way, the effect of the CVO swap remains local, as the edges of the above portion of the BBDD still point to the same logical function.

A variable swap $i \rightleftharpoons i + 1$ involves three CVO levels ($PV_{i+2} = w$, $SV_{i+2} = x$), ($PV_{i+1} = x$, $SV_{i+1} = y$) and ($PV_i = y$, $SV_i = z$). The level $i+2$ must be considered as it contains in SV the variable x, which is the PV swapped at level $i + 1$. If no level

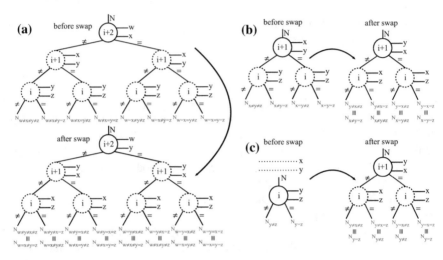

Fig. 2.6 Variable swap $i \rightleftharpoons i+1$ involving the CVO levels $(PV_{i+2} = w, SV_{i+2} = x), (PV_{i+1} = x, SV_{i+1} = y)$ and $(PV_i = y, SV_i = z)$. Effect on nodes at level $i + 2$ (**a**) $i + 1$ (**b**) and i (**c**)

$i + 2$ exists ($i + 1$ is the top level), the related operations are simply skipped. In the most general case, each node at level $i+2$, $i+1$ and i has 8, 4 and 2 possible children on the portion of BBDD below level i. Some of them may be identical, following to reduction rules **R1–4**, or complemented, deriving by the $\neq$-edges attributes in their path. Figure 2.6 depicts the different cases for a general node N located at level $i +2$, $i + 1$ or i, with all their possible children. After the swap $i \rightleftharpoons i + 1$, the order of comparisons $w \star x \star y \star z$ is changed to $w \star y \star x \star z$ and the children of N must be rearranged consequently ($\star \in \{=, \neq\}$). Using the *transitive property of equality and congruence* in the binary domain, it is possible to remap $w \star x \star y \star z$ into $w \star y \star x \star z$ as:

$$\begin{aligned} \star \in \{=, \neq\}, \quad & \bar{\star} : \{=, \neq\} \rightarrow \{\neq, =\} \\ (w \star_{i+2} x = y \star_i z) \rightarrow & (w \star_{i+2} y = x \star_i z) \\ (w \star_{i+2} x \neq y \star_i z) \rightarrow & (w \bar{\star}_{i+2} y \neq x \bar{\star}_i z) \end{aligned} \qquad (2.7)$$

Following remapping rules in Eq. 2.7, the children for N can be repositioned coherently with the variable swap. In Fig. 2.6, the actual children rearrangement after variable swap is shown.

In a bottom-up approach, it is possible to assemble back the swapped levels, while intrinsically respecting reduction rules **R1–4**, thanks to the *unique* table *strong canonical form*.

BBDD Reordering based on CVO Swap: Using the previously introduced CVO swap theory, global BBDD re-ordering can be carried out in different fashions. A popular approach for BDDs is the sifting algorithm presented in [47]. As its formulation is quite general, it happens to be advantageous also for BBDDs. Its BBDD implementation works as follows: Let n be the number of variables in the initial order π. Each variable π_i is considered in succession and the influence of the other variables is locally neglected. Swap operations are performed to move π_i in all n potential positions in the CVO. The best BBDD size encountered is remembered and its π_i position in the CVO is restored at the end of the variable processing. This procedure is repeated for all variables. It follows that BBDD sifting requires $O(n^2)$ swap operations.

Evolutions of the original sifting algorithm range between grouping of variables [49], simulated annealing techniques [50], genetic algorithms [51] and others. All of them are in principle applicable to BBDD reordering. In any of its flavors, BBDD reordering can be applied to a previously built BBDD or dynamically during construction. Usually, the latter strategy produces better results as it permits a tighter control of the BBDD size.

2.4 BBDD Representation: Theoretical and Experimental Results

In this section, we first show some theoretical properties for BBDDs, regarding the representation of majority and adder functions. Then, we present experimental results for BBDD representation of MCNC and HDL benchmarks, accomplished using the introduced BBDD software package.

2.4.1 Theoretical Results

Majority and adder functions are essential in many digital designs. Consequently, their efficient representation has been widely studied with state-of-art decision diagrams. We study hereafter the size for majority and adders with BBDDs and we compare these results with their known BDD size.

2.4.1.1 Majority Function

In Boolean logic, the majority function has an odd number n of inputs and an unique output. The output assumes the most frequent Boolean value among the inputs. With BBDDs, the MAJ_n function has a hierarchical structure. In Fig. 2.7, the BBDD for MAJ_7 is depicted, highlighting the hierarchical inclusion of MAJ_5 and MAJ_3. The key concepts enabling this hierarchical structure are:

(**M1**) $\neq$-edges reduce MAJ_n to MAJ_{n-2}: when two inputs assume opposite Boolean values they do not affect the majority voting decision.

(**M2**) $\lceil n/2 \rceil$ consecutive $=$-edges fully-determine MAJ_n voting decision: if $\lceil n/2 \rceil$ over n (odd) inputs have the same Boolean value, then this is the majority voting decision value.

Fig. 2.7 BBDD for the
7-input majority function.
The inclusion of MAJ$_5$ and
MAJ$_3$ functions is illustrated.
Grey nodes are nodes with
inverted children due to n to
$n-2$ majority reduction

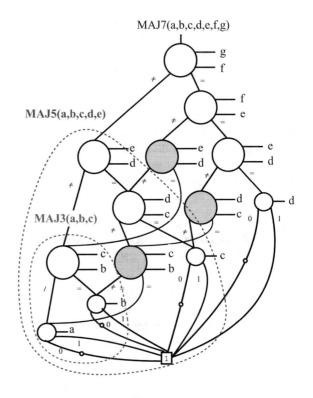

The M1 condition traduces in connecting $\neq$-edges to the BBDD structure for MAJ_{n-2}, or to local duplicated nodes with inverted children (see grey nodes in Fig. 2.7).

The M2 condition implies $\lceil n/2 \rceil$ consecutive BBDD nodes cascaded through $=$-edges.

Note that the variable order is not affecting the BBDD structure for a MAJ function as its behavior is invariant under input permutations [22].

Theorem 2.3 *A BBDD for the majority function of n (odd) variables has $\frac{1}{4}(n^2 + 7)$ nodes.*

Proof The M2 condition for MAJ_n requires $n-1$ nodes while the M1 condition needs the BBDD structure for MAJ_{n-2}. Consequently, the number of BBDD nodes is $|MAJ_n| = |MAJ_{n-2}| + n - 1$ with $|MAJ_3| = 4$ (including the sink node) as boundary condition. This is a non-homogeneous recurrence relation. Linear algebra methods [27] can solve such recurrence equation. The closed-form solution is $|MAJ_n| = \frac{1}{4}(n^2 + 7)$. ∎

Note that with standard BDDs, the number of nodes is $|MAJ_n| = \lceil \frac{n}{2} \rceil (n - \lceil \frac{n}{2} \rceil + 1) + 1$ [22]. It follows that BBDDs are always more compact than BDDs for majority,

Fig. 2.8 Full adder function
with BBDDs, variable order
$\pi = (a, b, cin)$

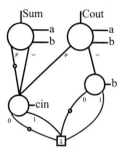

e.g., the BBDD for the 89-inputs majority function has 1982 nodes while its BDD counterpart has 2026 nodes. These values, and the law of *Theorem* 2.3, have been verified experimentally.

2.4.1.2   Adder Function

In Boolean logic, a n-bit adder is a function computing the addition of two n-bit binary numbers. In many logic circuits, a n-bit adder is represented as n cascaded 1-bit adders. A 1-bit binary adder, commonly called full adder, is a 3-input 2-output Boolean function described as $Sum = a \oplus b \oplus cin$ and $Cout = MAJ(a, b, cin)$. The BBDD for the full adder is depicted by Fig. 2.8.

With BBDDs, the 1-bit adder cascading concept can be naturally extended and leads to a compact representation for a general n-bit adder.

In Fig. 2.9, the BBDD of a 3-bit binary adder $(a + b)$, with $a = (a_2, a_1, a_0)$ and $b = (b_2, b_1, b_0)$, employing variable order $\pi = (a_2, b_2, a_1, b_1, a_0, b_0)$, is shown.

Theorem 2.4 *A BBDD for the n-bit binary adder function has $3n + 1$ nodes when the variable order $\pi = (a_{n-1}, b_{n-1}, a_{n-2}, b_{n-2}, \ldots, a_0, b_0)$ is imposed.*

Proof The proof follows by induction over the number of bit n and expanding the structure in Fig. 2.9. ∎

Note that the BDD counterpart for n-bit adders (best) ordered with
$$\pi = (a_{n-1}, b_{n-1}, a_{n-2}, b_{n-2}, \ldots, a_0, b_0)$$
has $5n + 2$ nodes [22]. For n-bit adders, BBDDs save about 40 % of the nodes compared to BDDs. These results, and the law of *Theorem* 2.4, have been verified experimentally.

2.4.2   Experimental Results

The manipulation and construction techniques described in Sect. 2.3.5 are implemented in a BBDD software package [19] using C programming language. Such

Fig. 2.9 BBDD for the 3-bit
binary adder function,
variable order
$\pi = (a_2, b_2, a_1, b_1, a_0, b_0)$

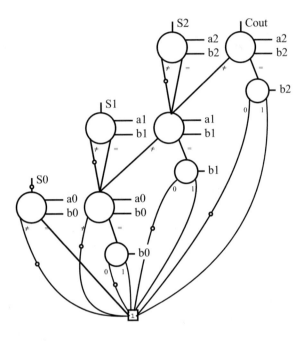

package currently counts about 8k lines of code. For the sake of comparison, we
consider CUDD [10] (manipulation package for BDDs) and *puma* [11] (manipulation package for KFDDs). We examine three categories of benchmarks: (i) MCNC
suite, (ii) portion of Open Cores designs and (iii) arithmetic HDL benchmarks. CUDD
and *puma* packages read BLIF format files while the BBDD package reads a Verilog
flattened onto primitive Boolean operations. The appropriate format conversion is
accomplished using ABC synthesis tool [17]. For all packages, dynamic reordering
during construction is enabled and based on the original sifting algorithm [47]. For
puma, also the choice of the most convenient decomposition type is enabled. The
machine running the experiments is a Xeon X5650 24-GB RAM machine. We verified with *Synopsys Formality* commercial tool the correctness of the BBDD output,
which is also in Verilog format.

   Table 2.1 shows the experimental results for the three packages. Note that the sizes
and runtime reported derives from heuristic techniques, so better results may exist.
Therefore, the following values provide an indication about the practical efficiency
of each decision diagram but do not give the means to determine if any of them is
globally superior to the others.

   MCNC Benchmarks: For large MCNC benchmarks, we report that BBDDs have
an average size 33.5 and 12.2% smaller than BDDs and KFDDs, respectively.
Regarding the runtime, the BBDD is 1.4× and 1.5× faster than CUDD and *puma*,
respectively. By handling two variables per time, BBDDs unlock new compact representation opportunities, not apparent with BDDs or KFDDs. Such size reduction
is responsible for the average runtime reduction. However, the general runtime for a

Table 2.1 Experimental results for DD construction using BBDDs, BDDs and KFDDs

Benchmarks	Inputs	Outputs	Wires	BBDD		CUDD (BDD)		puma (KFDD)	
				Node Count	Runtime (s)	Node Count	Runtime (s)	Node Count	Runtime (s)
MCNC benchmarks									
C1355	41	32	212	27701	1.22	68427	2.70	49785	8.32
C2670	233	64	825	29833	0.99	30329	0.88	36154	0.10
C499	41	32	656	32305	5.07	122019	5.60	49785	18.41
C1908	33	25	279	22410	0.53	18274	0.73	12716	0.08
C5315	178	123	1689	22263	1.03	42151	0.31	26658	0.57
C880	60	26	363	29362	0.40	22077	0.72	7567	0.03
C3540	50	22	1259	99471	8.93	93762	15.53	111324	0.73
C17	5	2	8	12	0.01	14	0.01	9	0.01
misex3	14	14	3268	766	0.08	870	0.02	1853	0.10
too_large	38	3	5398	1234	0.17	1318	0.26	6076	0.45
my_adder	33	17	98	166	0.09	620	0.11	456	0.21
Average	66.0	33.7	1277.7	**24138.4**	**1.7**	36351.0	2.4	27489.3	2.6
Combinational portions of open cores benchmarks									
custom-alu	37	17	193	2327	0.06	2442	0.01	2149	0.02
sin	35	64	2745	873	0.13	3771	0.12	1013	0.15
cosin	35	64	2652	851	0.10	3271	0.13	862	0.16
logsig	32	30	1317	1055	0.04	1571	0.09	1109	0.20
min-max	42	23	194	2658	0.40	2834	0.67	26736	0.76
h264-LUT	10	11	690	499	0.02	702	0.02	436	0.01
31-bit voter	31	1	367	242	0.01	257	0.01	256	0.01

(continued)

Table 2.1 (continued)

Benchmarks	Inputs	Outputs	Wires	BBDD		CUDD (BDD)		puma (KFDD)	
				Node Count	Runtime (s)	Node Count	Runtime (s)	Node Count	Runtime (s)
Combinational portions of open cores benchmarks									
ternary-adder	96	32	1064	366	0.32	8389	0.20	8389	0.20
max-weight	32	8	234	7505	0.15	7659	0.35	7610	0.55
cmul8	16	16	693	14374	0.55	12038	0.41	10979	0.21
fpu-norm	16	16	671	4209	0.12	7716	0.37	8608	0.32
Average	34.2	25.6	983.6	**3178.1**	**0.2**	4604.5	0.2	4022.2	0.2
Hard arithmetic benchmarks									
sqrt32	32	16	1248	223340	1145.53	11098772	3656.18	9256912	2548.92
hyperbola20	20	25	12802	126412	281.45	4522101	1805.20	4381924	2522.01
mult10x10	20	20	1123	123768	24.77	91192	15.74	91941	0.95
div16	32	32	3466	3675419	1428.87	7051263	7534.78	7842802	1583.22
Average	26.0	23.2	4659.7	**1.0e06**	**720.1**	5.6e06	3253.0	5.4e06	1671.3

decision diagram package is also dependent on the implementation maturity of the techniques supporting the construction. For this reason, there are benchmarks like C5315 where even if the final BBDD size is smaller than BDDs and KFDDs, its runtime is longer as compared to CUDD and *puma*, which have been highly optimized during years.

Open Cores Benchmarks: Combinational portions of Open Cores circuits are considered as representative for contemporary real-life designs. In this context, BBDDs have, on average, 30.9 and 20.9 % fewer nodes than BDDs and KFDDs, respectively. The average runtime is roughly the same for all packages. It appears that such benchmarks are easier than MCNC, having fairly small sizes and negligible runtime. To test the behavior of the packages at their limit we consider a separate class of hard circuits.

Arithmetic HDL Benchmarks: Traditional decision diagrams are known to face efficiency issues in the representation of arithmetic circuits, e.g., multipliers. We evaluate the behavior of the BBDD package in contrast to CUDD and *puma* for some of these hard benchmarks, i.e., a 10×10-bit multiplier, a 32-bit width square root unit, a 20-bit hyperbola and a 16-bit divisor. On average, BBDDs are about $5\times$ smaller than BDDs and KFDDs for such benchmarks. Moreover, the runtime of the BBDD package is $4.4\times$ faster than CUDD and *puma*. These results highlight that BBDDs have an enhanced capability to deal with arithmetic intensive circuits, thanks to the expressive power of the *biconditional expansion*. A theoretical study to determine the asymptotic bounds of BBDDs for these functions is ongoing.

2.5 BBDD-based Synthesis and Verification

This section showcases the interest of BBDDs in the automated design of digital circuits, for both standard CMOS and emerging silicon nanowire technology. We consider the application of BBDDs in logic synthesis and formal equivalence checking tasks for a real-life telecommunication circuit.

2.5.1 Logic Synthesis

The efficiency of logic synthesis is key to realize profitable commercial circuits. In most designs, critical components are arithmetic circuits for which traditional synthesis techniques do not produce highly optimized results. Indeed, arithmetic functions are not natively supported by conventional logic representation forms. Moreover, when intertwined with random logic, arithmetic portions are difficult to identify. Differently, BBDD nodes inherently act as two-variable comparators, a basis function for arithmetic operations. This feature opens the opportunity to restructure and identify arithmetic logic via BBDD representation.

We employ the BBDD package as front-end to a commercial synthesis tool. The BBDD restructuring is kept if it reduces the original representation complexity, i.e., the number of nodes and the number of logic levels. Starting from a simpler description, the synthesizer can reach higher levels of quality in the final circuit.

2.5.2 Formal Equivalence Checking

Formal equivalence checking task determines if two versions of a design are functionally equivalent. For combinational portions of a design, such task can be accomplished using canonical representation forms, e.g., decision diagrams, because equivalence test between two functions corresponds to a simple pointer comparison. BBDDs can speed up the verification of arithmetic intensive designs, as compared to traditional methods, thanks to their enhanced compactness.

We employ BBDDs to check the correctness of logic optimization methods by comparing an initial design with its optimized version.

2.5.3 Case Study: Design of an Iterative Product Code Decoder

To assess the effectiveness of BBDDs for the aforementioned applications, we design a real-life telecommunication circuit. We consider the *Iterative Decoder for Product Code* from Open Cores. The synthesis task is carried out using BBDD restructuring of arithmetic operations for each module, kept only if advantageous. The formal equivalence checking task is also carried out with BBDDs with the aim to speed-up the verification process. For the sake of comparison, we synthesized the same design without BBDD restructuring and we also verified it with BDDs in place of BBDDs.

As mentioned earlier, one compelling reason to study BBDDs is to provide a natural design abstraction for emerging technologies where the circuit primitive is a comparator, whose functionality is natively modeled by the *biconditional expansion*. For this reason, we target two different technologies: (i) a conventional CMOS 22-nm technology and (ii) an emerging controllable-polarity DG-SiNWFET 22-nm technology. A specific discussion for each technology is provided in the following subsections while general observations on the arithmetic restructuring are given hereafter.

The *Iterative Decoder for Product Code* consists of 8 main modules, among them 2 are sequential, one is the top entity, and 6 are potentially arithmetic intensive. We process the 6 arithmetic intensive modules and we keep the restructured circuits if their size and depth are decreased. For the sake of clarity, we show an example of restructuring for the circuit *bit_comparator*. Figure 2.10a depicts the logic network before processing and Fig. 2.10b illustrates the equivalent circuit after

Fig. 2.10 Representations for the bit_comparator circuit in [48] (inverters are bubbles in edges). **a** Original circuit **b** BBDD re-writing, reduced BDD nodes are omitted for the sake of illustration

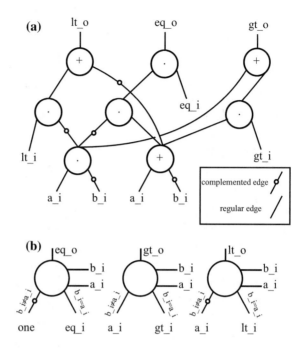

BBDD-restructuring. BDD nodes due to rule **R4** are omitted for simplicity. An advantage in both size and depth is reported. Table 2.2 shows the remaining results. BBDD-restructuring is favorable for all modules except *ext_val* that instead is more compact in its original version. The best obtained descriptions are finally given in input to the synthesis tool.

2.5.3.1 CMOS Technology

For CMOS technology, the design requirement is a clock period of 0.6 ns, hence a clock frequency of 1.66 GHz. The standard synthesis approach generates a negative slack of 0.12 ns, failing to meet the timing constraint. With BBDD-restructuring, instead, the timing constraint is met (slack of 0.00 ns), which corresponds to a critical path speedup of 1.2×. However, BBDD-restructuring induces a moderate area penalty of 9.6%.

2.5.3.2 Emerging DG-SiNWFET Technology

The controllable-polarity DG-SiNWFET technology features much more compact arithmetic (XOR, MAJ, etc.) gates than in CMOS, enabling faster and smaller implementation opportunities. For this reason, we set a tighter clock constraint

Table 2.2 Experimental results for BBDD-based design synthesis and verification

Case Study for Design and Verification: *Iterative Product Decoder*

Optimization via BBDD-rewriting

Logic Circuits	Type	I/O		BBDD-rewriting		Original		Gain
		Inputs	Outputs	Nodes	Levels	Nodes	Levels	
adder08_bit.vhd	Comb.	16	9	16	8	78	19	✓
bit_comparator.vhd	Comb.	5	3	3	1	8	3	✓
comparator_7bits.vhd	Comb.	14	3	21	7	58	14	✓
fulladder.vhd	Comb.	3	2	2	1	9	4	✓
ext_val.vhd	Comb.	16	8	674	16	173	29	x
twos_c_8bit.vhd	Comb.	8	8	20	8	29	8	✓
ser2par8bit.vhd	Seq.	11	64	–	–	–	–	–
product_code.vhd	Top	10	4	–	–	–	–	–

Synthesis in 22-nm CMOS technology—Clock period constraint: 0.6 ns (1.66 GHz)

		Inputs	Outputs	BBDD + Synthesis tool		Synthesis tool		Constraint met
				Area (μm²)	Slack (ns)	Area (μm²)	Slack (ns)	
product_code.vhd	Top	10	4	1291.03	0.00	1177.26	-0.12	✓

Synthesis in 22-nm DG-SiNWFET technology—Clock period constraint: 0.5 ns (2 GHz)

		Inputs	Outputs	BBDD + Synthesis tool		Synthesis tool		Constraint met
				Area (μm²)	Slack (ns)	Area (μm²)	Slack (ns)	
product_code.vhd	Top	10	4	1731.31	0.00	1673.78	-0.16	✓

Formal Equivalence Checking

		Inputs	Outputs	BBDD		CUDD (BDD)		Verification
				Nodes	Runtime	Nodes	Runtime	
product_code.vhd	Comb.	130	68	241530	185.11	227416	208.80	✓

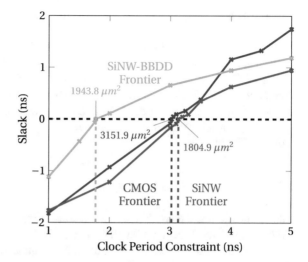

Fig. 2.11 Target *versus* obtained frequency curves and frequency frontiers for CMOS, SiNW-standard and SiNW-BBDD designs

than in CMOS, i.e., 0.5 ns corresponding to a clock frequency of 2 GHz. Direct synthesis of the design fails to reach such clock period with 0.16 ns of negative slack. With BBDD-restructuring, the desired clock period is instead reachable. For DG-SiNWFET technology, the benefit deriving from the use of BBDDs is even higher than in CMOS technology. Indeed, here BBDD-restructuring is capable to bridge a negative timing gap equivalent to 32 % of the overall desired clock period. For CMOS instead the same gap is just 20 %. This result confirms that BBDDs can further harness the expressive power of emerging technologies as compared to traditional synthesis techniques alone. Furthermore, the area penalty relative to BBDD-restructuring for DG-SiNWFET technology is decreased to only 3.3 %.

2.5.3.3 Post Place and Route Results

Using physical models for both CMOS and DG-SiNWFET technology, we also generated physical design results for the iterative product code decoder. In this set of experiments, the maximum clock period is determined by sweeping the clock constraint between 1 ns (1 GHz) and 5 ns (200 MHz) and repeating the implementation process. Figure 2.11 shows the post-*Place and Route* slack *versus* target clock constraint curves. Vertical lines highlight the clock constraint barriers for standard-SiNW (red), CMOS (blue) and BBDD-SiNW (green) designs. In the following, we report the shortest clock period achieved.

After place and route, the CMOS design reaches 331 MHz of clock frequency with area occupancy of 4271 μm^2 and EDP of 13.4 nJ.ns. The SiNWFET version, synthesized with plain design tools, has a slower clock frequency of 319 MHz and a larger EDP of 14.2 nJ.ns, but a lower area occupancy of 2473 μm^2. The final SiNWFET design, synthesized with BBDD-enhanced synthesis techniques, attains the fastest clock frequency of 565 MHz and the lowest EDP of 8.7 nJ.ns with a small $2643 \,\mu m^2$ of area occupancy.

If just using a standard synthesis tool suite, SiNWFET technology shows similar performances to CMOS, at the same technology node. This result alone would discard the SiNWFET technology from consideration because it brings no advantage as compared to CMOS. However, the use of BBDD abstraction and synthesis techniques enable a fair evaluation on the SiNWFETs technology, that is indeed capable of producing a faster and more energy efficient realization than CMOS for the *Iterative Product Code Decoder*.

2.5.3.4   Combinational Verification

The verification of the combinational portions of the *Iterative Decoder for Product Code* design took 185.11 s with BBDDs and 208.80 s and with traditional BDDs. The size of the two representations is roughly the same, thus the 12 % speed-up with BBDDs is accountable to the different growth profile of the decision diagrams during construction.

2.6   BBDDs as Native Design Abstraction for Nanotechnologies

BBDDs are the natural and native design abstraction for several emerging technologies where the circuit primitive is a comparator, rather than a switch. In this section, we test the efficacy of BBDDs in the synthesis of two emerging nanotechnologies other than the previously considered silicon nanowires: reversible logic and nanorelays. We start by introducing general notions on these two nanotechnologies in order to explain their primitive logic operation. Then, we show how the BBDD logic model fits and actually helps in exploiting at best the expressive power of the considered nanotechnologies.

Note that many other nanodevices may benefit from the presented biconditional synthesis methodologies [53, 54] however a precise evaluation of their performance is out of the scope of the current study.

2.6.1   Reversible Logic

The study of reversible logic has received significant research attention over the last few decades. This interest is motivated by the asymptotic zero power dissipation ideally achievable by reversible computation [57, 58]. Reversible logic finds application in a wide range of emerging technologies such as quantum computing [58], optical computing [59], superconducting devices and many others [60].

Fig. 2.12 Reversible circuit made of Toffoli, CNOT and NOT reversible gates

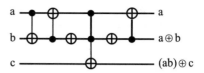

Reversible logic circuits are made of reversible logic gates [61]. Prominent reversible logic gates are, NOT gate: $Not(x) = x'$; CNOT gate: $CNOT(x, y) = (x, x \oplus y)$, which can be generalized with Tof_n gate with first $n-1$ variables acting as control lines: $Tof_n(x_1, x_2, \ldots, x_n, y) = (x_1, x_2, \ldots, x_n, (x_1 \cdot x_2 \cdot \ldots \cdot x_n) \oplus y)$. From a conceptual point of view, a CNOT gate is nothing but a Tof_n gate with $n = 1$. Analogously, a NOT gate is nothing but a Tof_n gate with $n = 0$. The Tof_n set of reversible logic gates form an universal gate library for realizing any reversible Boolean function. For the sake of clarity, we report in Fig. 2.12 an example of reversible circuit made of Toffoli reversible gates. We follow the established drawing convention of using the symbol $\oplus$ to denote the target line and solid black circles to indicate control connections for the gate. An $\oplus$ symbol with no control lines denotes a NOT gate.

Whether finally realized in one emerging technology or the other, reversible circuits must exploit at best the logic expressive power of reversible gates. Being the Toffoli gate the most known reversible gate, harnessing the biconditional connective embedded in its functionality is of paramount importance.

The efficiency of reversible circuits strongly depends on the capabilities of reversible synthesis techniques. Due to the inherent complexity of the reversible synthesis problem, several heuristics are proposed in the literature. Among those, the ones based on decision diagrams offer an attractive solution due to scalability and ability to trade-off diverse performance objectives.

Reversible circuit synthesis based on decision diagrams essentially consists of two phases. First, the generation of decision diagrams is geared towards efficient reversible circuit generation. This typically involves nodes minimization or other DD complexity metric reduction. Second, node-wise mapping is performed over a set of reversible gates.

The current standard for DD-based reversible synthesis uses binary decision diagrams generation via existing packages [10] and a custom node-wise mapping. However, standard BDDs do not match the intrinsic functionality of popular reversible gates that are comparator(XOR)-intensive. Instead, BBDDs are based on the biconditional expansion which natively models reversible XOR operations. In this way, BBDDs enable a more compact mapping into common reversible gates, such as Toffoli gates [55]. Figure 2.13 depicts the efficient mapping of a single BBDD node into reversible gates. The additional reversible gates w.r.t. a traditional BDD mapping are marked in gray. As one can notice, two extra gates are required. However, when comparing the functionality of BBDD nodes w.r.t. BDD nodes, it is apparent that more information is encoded into a single BBDD element. This is because the BBDD core expansion examines two variables per time rather than only one. Consequently,

Fig. 2.13 Reversible circuit
for a BBDD node [55]

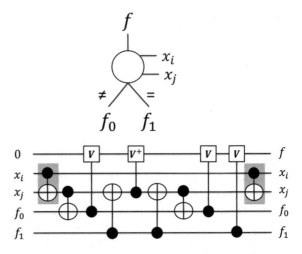

the node count reductions deriving from the use of BBDDs overcompensate the slight
increase in the direct mapping cost w.r.t. BDDs.

Our novel reversible synthesis flow uses BBDD logic representation and minimization using the package [19] and a final one-to-one mapping of BBDD nodes
as depicted by Fig. 2.13. As reference flow, we consider the traditional BDD-based
reversible synthesis approach. To validate the BBDD effectiveness, we run synthesis experiments over reversible benchmarks taken from the RevLib online library
[23]. In this context, we estimate the implementation cost using the *Quantum-Cost*
(QC) [55]. Table 2.3 shows the reversible synthesis results. Out of 26 benchmarks
functions studied, 20 reported improved QC and 13 reported improvement in QC
as well as line count. A closer study reveals that some benchmark functions, e.g.,
plus63mod4096, contain major contribution from non-linear sub-circuits, which are
represented in more compact form by BDD. This translates to better performance
in BDD-based synthesis. Nevertheless, future improvement in BBDD construction
heuristics may bridge also this gap.

These results provide a fair perspective on the efficacy of BBDDs in reversible
synthesis for emerging nanotechnologies.

2.6.2   NEMS

Nano-Electro-Mechanical Relay (NEMS), or simply nanorelays, are electrostatically
actuated mechanical switches [64]. The good properties of nanorelays are (i) very
low on-state intrinsic resistance (0.5Ω) and (ii) virtually infinitely large off-state
resistance [63]. On the other hand, the key hurdles of nanorelays are (i) long switching
time (hundreds of nanoseconds), (ii) relatively short lifetime (10^8 switching cycles)

Table 2.3 Results for reversible circuit synthesis using BBDDs versus traditional BDDs

Benchmark		BDD			BBDD			Improvement (%)	
Name	I/O	Line	QC	Runtime (s)	Line	QC	Runtime (s)	Line	QC
4mod5_8	4/1	7	24	<0.01	6	10	0.01	14.28	58.33
decod24_10	2/4	6	27	<0.01	6	23	0.02	0	14.81
mini-alu_84	4/2	10	60	<0.01	8	42	0.03	20.00	30.00
alu_9	5/1	7	29	0.01	7	25	0.02	0	13.79
rd53_68	5/3	13	98	<0.01	13	81	0.03	0	17.34
mod5adder_66	6/6	32	292	<0.01	32	269	0.05	0	7.76
rd73_69	7/3	13	217	<0.01	15	117	0.04	−15.38	46.08
rd84_70	8/4	34	304	<0.01	31	256	0.04	8.82	15.79
sym6_63	6/1	14	93	<0.01	11	49	0.02	21.43	47.31
sym9_71	9/1	27	206	<0.01	22	124	0.06	18.52	39.81
cycle10_2_61	12/12	39	202	0.09	25	183	0.03	35.89	9.41
cordic	23/2	52	325	0.06	50	222	0.02	3.84	31.69
bw	5/28	87	943	0.11	78	645	0.03	10.35	31.6
apex2	39/3	498	5922	0.24	744	5242	9.3	−49.39	11.48
seq	41/35	1617	19632	1.14	2440	18366	27.78	−50.89	6.45
spla	16/46	489	5925	0.10	788	5315	1.16	−61.15	10.3
ex5p	8/63	206	1843	0.24	251	1682	1.1	−21.85	8.74
e64	65/65	195	907	0.04	192	826	1.14	1.54	8.93
ham7_29	7/7	21	141	<0.01	18	153	0.03	14.29	−8.51
ham15_30	15/15	45	309	0.25	43	573	0.06	4.44	−85.44
hwb5_13	5/5	28	276	0.01	30	238	0.02	−7.14	13.77

(continued)

Table 2.3 (continued)

Benchmark		BDD			BBDD			Improvement (%)	
Name	I/O	Line	QC	Runtime (s)	Line	QC	Runtime (s)	Line	QC
hwb6_14	6/6	46	507	<0.01	49	488	0.06	−6.52	3.75
hwb7_15	7/7	73	909	<0.01	102	978	0.12	−39.73	−7.59
hwb8_64	8/8	112	1461	0.01	189	1831	0.35	−68.75	−25.33
plus63mod4096_79	12/12	23	89	0.08	28	186	0.16	−21.74	−108.99
plus127mod8192_78	13/13	25	98	0.21	31	210	0.02	−24.00	−114.28

and (iii) limited scalability of minimum feature size [62, 63]. Nanorelays can be fabricated by top-down approaches using conventional lithography techniques or bottom-up approaches using carbon nanotubes or nanowire beams [63].

Nanorelays are a promising alternative to CMOS for ultralow-power systems [62–66] where their ideally zero leakage current (consequence of the large off-resistance) is a key feature to be harnessed.

Different nanorelay structures for logic have been proposed in the literature. Most of them are based on electrostatic actuation and they implement different switching (logic) functions depending on their number of terminals and device geometry. Mechanical contacts (connections) are enforced via electric fields between the various terminals. Two-terminals (2T) and three-terminals (3T) nanorelays are simple devices useful to solve preliminary process challenges. Trading off simplicity for functionality, four-terminals (4T) and six-terminals (6T) nanorelays are more expressive and desirable for compact logic implementations.

In [68], a 4T NEM relay is proposed consisting of a movable poly-SiGe gate structure suspended above the tungsten body, drain, and source electrodes. Figure 2.14 shows the 4 T relay conceptual structure and a fabrication microphotograph. When a voltage is applied between the gate structure and the body electrode a corresponding electric field arises and the relay is turned on by the channel coming into contact with the source and drain electrodes.

In [67], a 6T NEM relay is realized by adding an extra body (*Body2*) and an extra source (*Source2*) contacts to the previous 4T NEM relay. Figure 2.15 shows the 6 T relay conceptual structure and a fabrication microphotograph. The two body contacts are designed to be biased by opposite voltages. Either *Source1* or *Source2* to *Drain* connection is controlled by the gate to body positive or negative voltage and its corresponding electric field polarity.

Because of the electrostatic forces among the different terminals, both 4 T and 6T NEM relay naturally acts as a logic multiplexer driven by a bit comparator.

In this study, we focus on 6T NEM relays. To assess the potential of nanorelays in VLSI, a BDD-based synthesis flow has been presented in [67]. It first partitions a design in sub-blocks and then creates BDDs for those sub-blocks. For each local BDD, a one-to-one mapping strategy generates a netlist of nanorelays implementing the target logic function. Indeed, the functionality of each BDD node can be realized by a single nanorelay device. We consider this as the reference design flow for nanorelays.

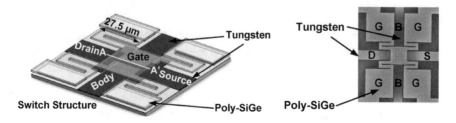

Fig. 2.14 Four-terminals nanorelay structure and fabrication image from [68]

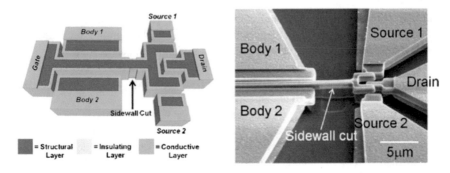

Fig. 2.15 Six-terminals nanorelay structure and fabrication image from [67]

Table 2.4 Total number of relays, the number of relays on the critical path, and ratios compared to [67] (MCNC benchmark circuits)

Circuit name	Relays	Levels	R. Ratio [67]	L. Ratio [67]
alu4	599	14	0.77	1.00
apex4	992	8	0.90	0.89
des	3130	18	0.78	1.00
ex1010	1047	10	0.94	0.91
ex5p	283	8	0.92	1.00
misex3	846	14	1.29	1.00
pdc	865	14	0.35	0.88
spla	691	16	0.82	1.00
8-b adder	28	9	0.19	0.53
16-b adder	56	17	0.31	0.52
8 × 8 multiplier	14094	16	1.05	1.00
Average	2057.36	13.09	0.76	0.88

From the analysis we performed above, we know that nanorelays can implement much more complex Boolean functions than just 2:1 multiplexers. Indeed, the functionality of these nanorelays is naturally modeled by a BBDD node. For this reason, we propose a novel design flow based on BBDDs to take full advantage of the nanorelays expressive power. Analogously to the BDD design flow, the design is first pre-partitioned if necessary. Then, local BBDDs are built and each BBDD node is mapped into a single nanorelay device.

We first test the BBDD-design flow against the MCNC benchmark suite. Table 2.4 shows the number of relays and the number of relays on the critical path. It compares these numbers with the corresponding numbers in [67] and shows the BBDD to BDD ratio for the different benchmark circuits. We also provide the ratios for the number of relays on the critical path. The BBDD design flow results in an average reduction in NEM relays of 24 %. This is due to the compactness of the BBDD representation relative to the BDD representation. Since BBDDs require less nodes

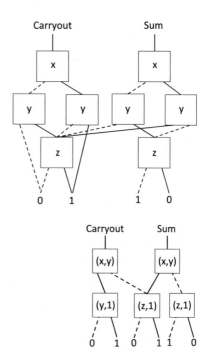

Fig. 2.16 Nanorelay implementation of a full-adder using a BDD-based design approach [67]

Fig. 2.17 Nanorelay implementation of a full-adder using a BBDD-based design approach. *Dotted lines* represent ≠-edges and *solid lines* are =-edges

than BDDs, BBDD circuits require less NEM relays. Furthermore, the BBDD design flow enables us to obtain circuits with shorter critical paths. On average the critical path length is reduced by 12 %. This decrease in the critical paths is due to the BBDD reduction rules, which can be leveraged to decrease the height of BBDDs more than the reduction rules for their BDD counterparts.

We also compare the BBDD-approach in the case of synthesizing an 8×8 array multiplier. In [67] the BDD-based approach is tested on such a multiplier implemented using a carry-save adder tree followed by a ripple carry adder. We represent the same multiplier, but using BBDDs instead of BDDs. The main source of advantage here is that BBDDs represent more compactly full-adders and half-adders as compared to BDDs. Figures 2.16 and 2.17 depicts the nanorelay implementation of a full-adder using BDD and BBDD approaches, respectively. Each squared box in the figures represents a six-terminal nanorelay device. We show that this compact representation allows us to implement the multiplier using a smaller number of NEM relays. Table 2.5 shows the corresponding results. It is possible to see that the BBDD design flow requires a smaller number of relays. On average, reduce the number of relays is reduced by 36 % w.r.t. the BDD design flow. Furthermore, as the number of mechanical delays decreases, so does the ratio of the number of relays required by the BBDD representation versus the BDD representation.

These results show the impact of a dedicated logic abstraction to design a comparator-intrinsic nanotechnology, such as nanorelays.

Table 2.5 Comparison of BDD-based versus BBDD-based synthesis of an 8×8 array multiplier

Mech. Delays	BBDD-Relays	BDD-Relays	Ratio BBDD/BDD
2	2491	4129	0.60
3	1367	2186	0.63
4	647	875	0.74
5	434	590	0.74
6	407	533	0.76
Average	1069.2	1662.6	0.64

2.7 Summary

Following the trend to handle ever-larger designs, and in light of the rise of emerging technologies that natively implement new Boolean primitives, the study of innovative logic representation forms is extremely important. In this chapter, we proposed Biconditional BDDs, a new canonical representation form driven by the *biconditional expansion*. BBDDs implement an equality/inequality switching paradigm that enhances the expressive power of decision diagrams. Moreover, BBDDs natively models the functionality of emerging technologies where the circuit primitive is a comparator, rather than a simple switch. Employed in electronic design automation, BBDDs (i) push further the efficiency of traditional decision diagrams and (ii) unlock the potential of promising post-CMOS devices. Experimental results over different benchmark suites, demonstrated that BBDDs are frequently more compact than other decision diagrams, from $1.1\times$ to $5\times$, and are also built faster, from $1.4\times$ to $4.4\times$. Considering the synthesis of a telecommunication circuit, BBDDs advantageously restructure critical arithmetic operations. With a 22-nm CMOS technology, BBDD-restructuring shorten the critical path by 20% (14% post *place and route*). With an emerging 22-nm controllable-polarity DG-SiNWFET technology, BBDD-restructuring shrinks more the critical path by 32% (22% post *place and route*), thanks to the natural correspondence between device operation and logic representation. The formal verification of the optimized design is also accomplished using BBDDs in about 3 min, which is about 12% faster than with standard BDDs. Results on other two nanotechnologies, i.e., reversible logic and nanorelays, demonstrate that BBDDs are essential to permit a fair technology evaluation where the logic primitive is a binary comparator.

References

1. C.Y. Lee2, Representation of switching circuits by binary-decision programs. Bell Syst. Tech. J. **38**(4), 985–999 (1959)
2. S.B. Akers2, Binary decision diagrams. IEEE Trans. Comp. **100**(6), 509–516 (1978)

3. R.E. Bryant, Graph-based algorithms for Boolean function manipulation. IEEE Trans. Comput. **100**(8), 677–691 (1986)
4. C. Yang, M. Ciesielski, BDS: A BDD-based logic optimization system. IEEE Trans. CAD IC Syst. **21**(7): 866–876 (2002)
5. S. Malik et al., Logic verification using binary decision diagrams in a logic synthesis environment in *IEEE International Conference on CAD* (1988), pp. 6–9
6. M.S. Abadir et al., Functional test generation for digital circuits using binary decision diagrams. IEEE Trans. Comput. **100**(4), 375–379 (1986)
7. C. Scholl, R. Drechsler, B. Becker, Functional simulation using binary decision diagrams in *IEEE International Conference on CAD* (1997), pp. 8–12
8. U. Kebschull, W. Rosenstiel, E. Schubert, Multilevel logic synthesis based on functional decision diagrams, in *IEEE European Conference Design Automation* (1992), pp. 43–47
9. R. Drechsler et al., Ordered Kronecker functional decision diagrams-a data structure for representation and manipulation of Boolean functions. IEEE Trans. CAD IC Syst. **17**(10), 965–973 (1998)
10. *CUDD: CU Decision Diagram Package Release 2.5.0*. http://vlsi.colorado.edu/fabio/CUDD/cuddIntro.html
11. *Decision Diagram-Package PUMA*. http://ira.informatik.uni-freiburg.de/software/puma/pumamain.html
12. M. De Marchi et al., Polarity control in double-gate (gate-all-around vertically stacked silicon nanowire FETs, in *IEEE Electron Devices Meeting* (2012), pp. 8–14
13. Y. Lin et al., High-performance carbon nanotube field-effect transistor with tunable polarities. IEEE Trans. Nanotech. **4**(5), 481–489 (2005)
14. N. Harada et al., A polarity-controllable graphene inverter. Appl. Phys. Lett. **96**(1), 012102 (2010)
15. D. Lee2 et al., Combinational logic design using six-terminal NEM relays. IEEE Trans. CAD IC Syst. **32**(5), 653–666 (2013)
16. L. Amaru, P.-E. Gaillardon, S. Mitra, G. DeMicheli, New logic synthesis as nanotechnology enabler, in *Proceedings of the IEEE* (2015)
17. L. Amaru, P.-E. Gaillardon, G. De Micheli, Biconditional BDD: a novel canonical representation form targeting the synthesis of XOR-rich circuits, in Design Automation and Test in Europe (2013), pp. 1014–1017
18. L. Amaru, P.-E. Gaillardon, G. De Micheli, An efficient manipulation package for biconditional binary decision diagrams, in *Design Automation and Test in Europe* (2014), pp. 296–301
19. *BBDD package*. http://lsi.epfl.ch/BBDD
20. L. Kathleen, *Logic and Boolean Algebra*, Barrons Educational Series (1979)
21. G. De Micheli, *Synthesis and Optimization of Digital Circuits* (McGraw-Hill, New York, 1994)
22. I. Wegener, *Branching Programs and Binary Decision Diagrams: Theory and Applications*, vol. 4 (SIAM, Philadelphia, 2000)
23. M. Kreuzer, L. Robbiano, *Computational Commutative Algebra*, vol. 1 (Springer, Berlin, 2005)
24. R.E. Bryant, On the complexity of VLSI implementations and graph representations of boolean functions with application to integer multiplication. IEEE Trans. Comput. **40**(2), 205–213 (1991)
25. J. Gergov, C. Meinel, Mod-2-OBDDs A data structure that generalizes EXOR-sum-of-products and ordered binary decision diagrams. Form. Methods Syst. Des. **8**(3), 273–282 (1996)
26. B. Bollig, Improving the variable ordering of OBDDs is NP-complete. IEEE Trans. Comput. **45**(9), 993–1002 (1996)
27. T. Koshy, *Discrete Mathematics with Applications* (Academic Press, Cambridge, 2004)
28. T.S. Czajkowski, S.D. Brown, Functionally linear decomposition and synthesis of logic circuits for FPGAs. IIEEE Trans. CAD IC Syst. **27**(12), 2236–2249 (2008)
29. J.F. Groote, J. Van de Pol, *Equational Binary Decision Diagrams*, Logic for programming and automated reasoning (Springer, Heidelberg, 2000)
30. S. Minato, Zero-suppressed BDDs for set manipulation in combinatorial problems, in *IEEE Conference on Design Automation (DAC)* (1993), pp. 272–277

31. C. Meinel, F. Somenzi, T. Theobald, Linear sifting of decision diagrams, in *IEEE Conference on Design Automation (DAC)* (1997), pp. 202–207
32. W. Gunther, R. Drechsler, BDD minimization by linear transformations. Adv. Comput. Syst. 525–532 (1998)
33. M. Fujita, Y. Kukimoto, R. Brayton, BDD minimization by truth table permutation, in *IEEE International Symposium on CAS* (1996), pp. 596–599
34. E.M. Clarke, M. Fujita, X. Zhao, Hybrid decision diagrams, in *IEEE International Conference on CAD* (1995), pp. 159–163
35. E.I. Goldberg, Y. Kukimoto, R.K. Brayton, Canonical TBDD's and their application to combinational verification, in *ACM/IEEE International Workshop on Logic Synthesis* (1997)
36. U. Kebschull, W. Rosenstiel, Efficient graph-based computation and manipulation of functional decision diagrams, in *IEEE Euro Conference on Design Automation* (1993), pp. 278–282
37. J.E. Rice, Making a choice between BDDs and FDDs, in *ACM/IEEE International Workshop on Logic Synthesis* (2005)
38. R. Drechsler, Ordered Kronecker Functional Decision Diagrams und ihre Anwndung, Ph.D. Thesis, 1996
39. S. Grygiel, M.A. Perkowski, New compact representation of multiple-valued functions, relations, and non-deterministic state machines, in *IEEE International Conference on Computer Design* (1998), pp. 168–174
40. A. Srinivasan, T. Kam, S. Malik, R. Brayton, Algorithms for Discrete Function Manipulation, in *IEEE International Conference on CAD* (1990), pp. 92–95
41. S. Minato et al., Shared BDD with attributed edges for efficient boolean function manipulation, in *IEEE Conference on Design Automation (DAC)* (1990), pp. 52–57
42. B. Becker, R. Drechsler, How many decomposition types do we need?, in *IEEE Euro Conference on Design Automation* (1995), pp. 438–442
43. B. Becker, R. Drechsler, M. Theobald, On the expressive power of OKFDDs. Form. Methods Syst. Des. **11**(1), 5–21 (1997)
44. R. Drechsler, B. Becker, *Binary Decision Diagrams: Theory and Implementation* (Kluwer Academic Publisher, The Netherlands, 1998)
45. P. Tarau, *Pairing Functions, Boolean Evaluation and Binary Decision Diagrams*, arxiv preprint arXiv:0808.0555 (2008)
46. K.S. Brace, R.L. Rudell, R.E. Bryant, Efficient implementation of a BDD package, in *IEEE Conference on Design Automation (DAC)* (1990), pp. 40–45
47. R. Rudell, Dynamic variable ordering for ordered binary decision diagrams, in *IEEE International Conference on CAD* (1993), pp. 42–47
48. *An iterative decoder for Product Code—from Open Cores*: http://opencores.org/project, product_code_iterative_decoder
49. S. Panda, F. Somenzi, Who are the variables in your neighborhood, in *IEEE International Conference on CAD* (1995), pp. 74–77
50. B. Bollig et al., Simulated annealing to improve variable orderings for OBDDs, in *ACM/IEEE International Workshop on Logic Synthesis* (1995)
51. R. Drechsler et al., A genetic algorithm for variable ordering of OBDDs, in *ACM/IEEE International Workshop on Logic Synthesis* (1995)
52. ABC synthesis tool - available online
53. J. Hagenauer, E. Offer, L. Papke, Iterative decoding of binary block and convolutional codes. IEEE Trans. Inf. Theory **42**(2), 429–445 (1996)
54. A. Picart, R. Pyndiah, Adapted iterative decoding of product codes, in *Global Telecommunications Conference, 1999. GLOBECOM'99*, vol. 5 (IEEE, New York, 1999)
55. A. Chattopadyay, et al., Reversible logic synthesis via biconditional binary decision diagrams, in *Proceedings of the ISMVL 15*
56. RevLib is an online resource for benchmarks within the domain of reversible and quantum circuit design. http://www.revlib.org
57. C.H. Bennett, Logical reversibility of computation. IBM J. Res. Dev. **17**(6), 525532 (1973)

58. M. Nielsen, I. Chuang, *Quantum Computation and Quantum Information* (Cambridge University Press, Cambridge, 2000)
59. R. Cuykendall, D.R. Andersen, Reversible optical computing circuits. Opt. Lett. **12**(7), 542544 (1987)
60. R.C. Merkle, Reversible electronic logic using switches, in *Nanotechnology*, vol. 4 (1993), p. 2140
61. A. Barenco et al., Elementary gates for quantum computation, in *Physical Review* (1995)
62. O. Loh, H. Espinosa, Nanoelectromechanical contact switches. Nat. Nanotechnol. **7**(5), 283–295 (2012)
63. *Nano-Electro-Mechanical Switches*, ITRS, white paper (2008)
64. V. Pott et al., Mechanical computing redux: relays for integrated circuit applications. Proc. IEEE **98**(12), 2076–2094 (2010)
65. Sharma, P., Perruisseau-Carrier, J., Moldovan, C., Ionescu, A. Electromagnetic performance of RF NEMS graphene capacitive switches. IEEE Trans. Nanotech. (2014)
66. Dana Weinstein, Sunil A. Bhave, The resonant body transistor. Nano Lett. **10**(4), 1234–1237 (2010)
67. D. Lee et al., Combinational logic design using six-terminal NEM relays. IEEE Trans. Comput. Aided Des. Integr. Circuits Syst. **32**(5), 653–666 (2013)
68. M. Spencer et al., Demonstration of integrated micro-electro-mechanical relay circuits for VLSI applications. IEEE J. Solid State Circuits **46**(1), 308–320 (2011)

Chapter 3
Majority Logic

In this chapter, we propose a paradigm shift in representing and optimizing logic by using only majority (MAJ) and inversion (INV) functions as basic operations. We represent logic functions by *Majority-Inverter Graph* (MIG): a directed acyclic graph consisting of three-input majority nodes and regular/complemented edges. We optimize MIGs via a new Boolean algebra, based exclusively on majority and inversion operations, that we formally axiomatize in this work. As a complement to MIG algebraic optimization, we develop powerful Boolean methods exploiting global properties of MIGs, such as bit-error masking. MIG algebraic and Boolean methods together attain very high optimization quality. For example, when targeting depth reduction our MIG optimizer, *MIGhty*, transforms a ripple carry adder into a carry look-ahead like structure. Considering the set of IWLS'05 benchmarks, *MIGhty* enables a 7 % depth reduction in LUT-6 circuits mapped by ABC academic tool, while also reducing size and power activity, with respect to similar *And-Inverter Graph* (AIG) optimization. Focusing instead on arithmetic intensive benchmarks, *MIGhty* enables a 16 % depth reduction in LUT-6 circuits mapped by ABC academic tool, again with respect to similar AIG optimization. Employed as front-end to a delay-critical 22-nm ASIC flow (logic synthesis + physical design), *MIGhty* reduces the average delay/area/power by 13 %/4 %/3 %, respectively, over 31 academic and industrial benchmarks. We also demonstrate improvements in delay/area/power metrics by 10 %/10 %/5 % for a commercial FPGA flow. Furthermore, MIGs are the natural design abstraction for emerging nanotechnologies whose logic primitive is a majority voter. Results on two of these nanotechnologies, i.e., spin-wave devices and resistive RAM, show the efficacy of MIG-based synthesis. Finally, we extend the majority logic axiomatization from 3 to n inputs, with n being odd.

© Springer International Publishing Switzerland 2017
L.G. Amaru, *New Data Structures and Algorithms for Logic
Synthesis and Verification*, DOI 10.1007/978-3-319-43174-1_3

3.1 Introduction

Nowadays, *Electronic Design Automation* (EDA) tools are challenged by design goals at the frontier of what is achievable in advanced technologies. In this scenario, extending the optimization capabilities of logic synthesis tools is of paramount importance.

In this chapter, we propose a paradigm shift in representing and optimizing logic, by using only majority (MAJ) and inversion (INV) as basic operations. We use the terms inversion and complementation interchangeably. We focus on majority functions because they lie at the core of Boolean function classification [1]. Figure 3.1 depicts the Boolean function classification presented in [1] together with the hierarchical inclusion among notable classes. We give an informal description of the main classes hereafter. A monotone increasing (decreasing) function is a function that can be represented by a *Sum-Of-Products* (SOP) with (without) complemented literals. A unate function is a generalization of a monotone function. A function is unate if it can be represented by a SOP using either uncomplemented or complemented literals for each variable. A threshold function, with threshold k, evaluates to logic one on input vectors with k or more ones. All threshold functions are unate but necessarily monotone. A self-dual function is a function such that its output complementation is equivalent to its inputs complementation. Self-dual functions are not fully included by any of the previous classes. A majority function evaluates to logic one on input vectors having more ones than zeros. Majority functions are threshold, unate, monotone increasing and self-dual at the same time. Together with inversion, majority can express all Boolean functions. Note that minority gates, which represent complemented majority functions, are common in VLSI because they natively implement carry functions.

Based on these primitives, we present in this work the *Majority-Inverter Graph* (MIG), a logic representation structure consisting of three-input majority nodes and regular/complemented edges. MIGs include any *AND/OR/Inverter Graphs* (AOIGs), containing also the popular AIGs [2]. To provide native manipulation of MIGs, we introduce a novel Boolean algebra, based exclusively on majority and inver-

Fig. 3.1 Relations among various functions extracted from [1]

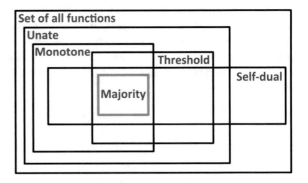

sion operations [3]. We define a set of five transformations forming a sound and complete axiomatic system. Using a sequence of these primitive axioms, it is possible to manipulate efficiently a MIG and reach all points in the representation space. We apply MIG algebra axioms locally, to design fast and efficient MIG algebraic optimization methods. We also exploit global properties of MIGs to design slower but very effective MIG Boolean optimization methods [4]. Specifically, we take advantage of the error masking property of majority operators. By selectively inserting logic errors in a MIG, successively masked by majority nodes, we enable strong simplifications in logic networks. MIG algebraic and Boolean methods together attain very high optimization quality. For example when targeting depth reduction, our MIG optimizer, *MIGhty*, transforms a ripple carry structure into a carry look-ahead like one. Considering the set of IWLS'05 benchmarks, *MIGhty* enables a 7% depth reduction in LUT-6 circuits mapped by ABC [2] while also reducing size and power activity, with respect to similar AIG optimization. Focusing on arithmetic intensive benchmarks, *MIGhty* enables a 16% depth reduction in LUT-6 circuits, again with respect to similar AIG optimization. Employed as front-end to a delay-critical 22-nm ASIC flow (logic synthesis + physical design), *MIGhty* reduces the average delay/area/power by 13%/4%/3%, respectively, over 31 academic and industrial benchmarks, as compared to a leading commercial ASIC flow. We demonstrate improvements in delay/area/power metrics by 10%/10%/5% for a commercial 28-nm FPGA flow. MIGs are also the native logic abstraction for circuit design in nanotechnologies whose logic primitive is a majority voter.

The remainder of this chapter is organized as follows. Section 3.2 gives background on logic representation and optimization. Section 3.3 presents MIGs with their properties and associated Boolean algebra. Section 3.4 proposes MIG algebraic optimization methods and Sect. 3.5 describes MIG Boolean optimization methods. Section 3.6 shows experimental results for MIG-based optimization and compares them to the state-of-the-art academic tools. Section 3.6 also shows results for MIG-based optimization employed as front-end to commercial ASIC and FPGA design flows. Section 3.7 gives a vision on future nanotechnologies design via MIGs. Section 3.8 extends the theory results from 3 to arbitrary n-ary majority operators, with n odd. Section 3.9 concludes the chapter.

3.2 Background and Motivation

This section presents first a background on logic representation and optimization for logic synthesis. Then, it introduces the necessary notations and definitions for this work.

3.2.1 Logic Representation

The (efficient) way logic functions are represented in EDA tools is key to design efficient hardware. On the one hand, logic representations aim at having the fewest number of primitive elements (literals, sum-of-product terms, nodes in a logic network, etc.) in order to (i) have small memory footprint and (ii) be covered by as few library elements as possible. On the other hand, logic representation forms must be simple enough to manipulate. This may require having a larger number of primitive elements but with simpler manipulation laws. The choice of a computer data-structure is a trade-off between compactness and manipulation easiness.

In the early days of EDA, the standard representation form for logic was the *Sum Of Product* (SOP) form, i.e., a disjunction (OR) of conjunctions (AND) made of literals [5].This standard was driven by PLA technology whose functionality is naturally modeled by a SOP [6]. Other two-level forms, such as product-of-sums or EX-SOP, have been studied at that time [2]. Two-level logic is compact for small sized functions but, beyond that size, it becomes too large to be efficiently mapped into silicon. Yet, two-level logic has been supported by efficient heuristic and exact optimization algorithms. With the advent of VLSI, the standard representation for logic moved from SOP to *Directed Acyclic Graphs* (DAGs) [1]. In a DAG-based logic representation, nodes correspond to logic functions (gates) and directed edges (wires) connect the nodes. Nodes' functions can be internally represented by SOPs leveraging the proven efficiency of two-level optimization. From a global perspective, general optimization procedures run on the entire DAG. While being potentially very compact, DAGs without bounds on the nodes' functionality do not support powerful logic optimization. This is because this kind of representation demands that optimization techniques deal with all possible types and sizes of functions which is impractical. Moreover, the cumulative memory footprint for each *functionally unbounded* node is potentially very large. Restricting the permissible node function types alleviates this issue. At the extreme case, one can focus on just one type of function per node and add complemented/regular attributes to the edges. Even though in principle, this restriction increases the representation size, in practice it unlocks better (smaller) representations because it supports more effective logic optimization simplifying a DAG. A notable example of DAG where all the nodes realize the same function is *Binary Decision Diagrams* (BDDs) [7]. In BDDs, nodes act as 2:1 multiplexers. We refer the reader to Sect. 2.2.1 for a complete background on BDDs. Another DAG where all nodes realize the same function is the *And-Inverter Graph* (AIG) [2, 8] where nodes act as two-inputs ANDs. AIGs can be optimized through traditional Boolean algebra axioms and derived theorems. Iterated over the whole AIG, local transformations produce very effective results and scale well with the size of the circuits. This means that, overall, AIGs can be made remarkably small through logic optimization. For this reason, AIG is one of the current representation standards for logic synthesis.

With the ever-increasing complexity of digital design, DAGs with restricted node functionality (ideally to one) provide a scalable approach to manipulate logic functions. In this scenario, choosing a node functionality is critical as it determines a representation compactness and manipulation easiness. In this work, we show that majority operators are excellent candidates for this role. While having an enhanced expressiveness with respect to traditional AND/ORs, majority operators also enable more capable optimization strategies leading to superior synthesis results.

3.2.2 Logic Optimization

Logic optimization consists of manipulating a logic representation structure in order to minimize some target metric. Usual optimization targets are size (number of nodes/elements), depth (maximum number of levels), interconnections (number of edges/nets), etc.

Logic optimization methods are closely coupled to the data structures they run on. In two-level logic representation (SOP), optimization aims at reducing the number of terms. ESPRESSO is the main optimization tool for SOP [6]. Its algorithms operate on SOP cubes and manipulate the ON-, OFF- and *Don't Care* (DC)-covers iteratively. In its default settings, ESPRESSO uses fast heuristics and does not guarantee to reach the global optimum. However, an exact optimization of two level logic is available (under the name of ESPRESSO-exact) and often run in a reasonable time. The exact two-level optimization is based on Quine-McCluskey algorithm [9]. Moving to DAG logic representation (also called multi-level logic), optimization aims at reducing graph size and depth or other accepted complexity metrics. There, DAG-based logic optimization methods are divided into two groups: Algebraic methods, which are fast and Boolean methods, which are slower but may achieve better results [10]. Traditional algebraic methods assume that DAG nodes are represented in SOP form and treat them as polynomials [11, 12]. Algebraic operations are selectively iterated over all DAG nodes, until no improvement is possible. Basic algebraic operations are extraction, decomposition, factoring, balancing and substitution [10, 13]. Their efficient runtime is enabled by theories of weak-division and kernel extraction. In contrast, Boolean methods do not treat the functions as polynomials but handle their true Boolean nature using Boolean identities as well as (global) don't cares (circuit flexibilities) to get a better solution [1, 10, 14–16]. Boolean division and substitution techniques trade off runtime for better minimization quality. Functional decomposition is another Boolean method which aims at representing the original function by means of simpler component functions. The first attempts at functional decomposition [17–19] make use of decomposition charts to find the best component functions. Since the decomposition charts grows exponentially with the number of variables these techniques are only applicable to small functions. A different, and more scalable, approach to functional decomposition is based on the BDD data structure. A particular class of BDD nodes, called dominator nodes, highlights advantageous functional decomposition points [20]. BDD decomposition can be applied recursively and is capable of exploiting optimization opportunities not visible by algebraic counterparts [20–22]. Recently, disjoint support decomposition has also been considered to optimize locally small functions and speedup logic manipulation [23, 24]. It is worth mentioning that the main difficulty in developing Boolean algorithms is due to the unrestricted space of choices. This makes more difficult to take good decisions during functional decomposition.

Advanced DAG optimization methodologies, and associated tools, use both algebraic and Boolean methods. When DAG nodes are restricted to just one function type, the optimization procedure can be made much more effective. This is because logic transformations are designed specifically to target the functionality of the chosen node. For example, in AIGs, logic transformations such as balancing, refactoring, and general rewriting are very effective. For example, balancing is based on the associativity axiom from traditional Boolean algebra [25, 26]. Refactoring operates on an AIG subgraph which is first collapsed into SOP and then factored out [12]. General rewriting conceptually includes balancing and refactoring. Its purpose is to replace AIG subgraphs with equivalent pre-computed AIG implementations that improve the number of nodes and levels [25]. By applying local, but powerful, transformations many times during AIG optimization it is possible to obtain very good result quality. The restriction to AIGs makes it easier to assess the intermediate quality and to develop the algorithms, but in general is more prone to local minimum. Nevertheless, Boolean methods can still complement AIG optimization to attain higher quality of results [14, 27].

In this chapter, we present a new representation form, based on majority and inversion, with its native Boolean algebra. We show algebraic and Boolean optimization techniques for this data structure unlocking new points in the design space.

Note that early attempts to majority logic have already been reported in the 60s [28–30], but, due to their inherent complexity, failed to gain momentum later on in automated synthesis. We address, in this chapter, the unique opportunity led by majority logic in a contemporary synthesis flow.

3.2.3   Notations and Definitions

We provide hereafter notations and definitions on Boolean algebra and logic networks.

3.2.3.1   Boolean Algebra

In the binary Boolean domain, the symbol $\mathbb{B}$ indicates the set of binary values $\{0, 1\}$, the symbols $\wedge$ and $\vee$ represent the conjunction (AND) and disjunction (OR) operators, the symbol $'$ represents the complementation (INV) operator and 0/1 are the false/true logic values. Alternative symbols for $\wedge$ and $\vee$ are $\cdot$ and $+$, respectively. The standard Boolean algebra (originally axiomatized by Huntington [31]) is a non-empty set $(\mathbb{B}, \cdot, +, ', 0, 1)$ subject to *identity, commutativity, distributivity, associativity* and *complement* axioms over $\cdot$, $+$ and $'$ [1]. For the sake of completeness, we report these basic axioms in Eq. 3.1. Such axioms will be used later on in this work for proving theorems.

This axiomatization for Boolean algebra is sound and complete [32]. Informally, it means that logic arguments, or formulas, proved by axioms in Δ (defined below in Eq. 3.1) are valid (soundness) and all true logic arguments are provable (completeness). We refer the reader to [32] for a more formal discussion on mathematical logic. In practical EDA applications, only sound and complete axiomatizations are of interest.

Other Boolean algebras exist, with different operators and axiomatizations, such as Robbins algebra, Freges algebra, Nicods algebra, etc. [32]. Binary Boolean algebras are the basis to operate on logic networks.

$$
\Delta \begin{cases}
\textbf{Identity} : \Delta.I \\
x + 0 = x \\
x \cdot 1 = x \\
\textbf{Commutativity} : \Delta.C \\
x \cdot y = y \cdot x \\
x + y = y + x \\
\textbf{Distributivity} : \Delta.D \\
x + (y \cdot z) = (x + y) \cdot (x + z) \\
x \cdot (y + z) = (x \cdot y) + (x \cdot z) \\
\textbf{Associativity} : \Delta.A \\
x \cdot (y \cdot z) = (x \cdot y) \cdot z \\
x + (y + z) = (x + y) + z \\
\textbf{Complement} : \Delta.Co \\
x + x' = 1 \\
x \cdot x' = 0
\end{cases} \tag{3.1}
$$

3.2.3.2   Logic Network

A logic network is a *Directed Acyclic Graph* (DAG) with nodes corresponding to logic functions and directed edges representing interconnection between the nodes. The direction of the edges follows the natural computation from inputs to outputs. The terms logic network, Boolean network, and logic circuit are used interchangeably in this paper. Two logic networks are said *equivalent* when they represent the same Boolean function. A logic network is said *irredundant* if no node can be removed without altering the Boolean function that it represents. A logic network is said *homogeneous* if each node represents the same logic function and has a fixed indegree, i.e., the number of incoming edges or fan-in. In a homogeneous logic network, edges can have a regular or complemented attribute. The depth of a node is the length of the longest path from any primary input variable to the node. The depth of a logic network is the largest depth among all the nodes. The size of a logic network is the number of its nodes.

3.2.3.3   Self-dual Function

A logic function $f(x, y, \ldots, z)$ is said to be *self-dual* if $f = f'(x', y', \ldots, z')$ [1]. By complementation, an equivalent *self-dual* formulation is $f' = f(x', y', \ldots, z')$. For example, the function $f = x'y'z' + x'yz + xy'z + xyz'$ is self-dual.

3.2.3.4   Majority Function

The n-input (n being odd) majority function M returns the logic value assumed by more than half of the inputs [1]. More formally, a majority function of n variables returns logic one if a number of input variables k over the total n, with $k \geq \lceil \frac{n}{2} \rceil$, are equal to logic one. For example, the three input majority function $M(x, y, z)$ is represented in terms of $\cdot, +$ by $(x \cdot y) + (x \cdot z) + (y \cdot z)$. Also $(x + y) \cdot (x + z) \cdot (y + z)$ is a valid representation for $M(x, y, z)$. The majority function is *self-dual* [1].

3.3   Majority-Inverter Graphs

In this section, we present MIGs and their representation properties. Then, we show a new Boolean algebra natively fitting the MIG data structure. Finally, we discuss the error masking capabilities of MIGs from an optimization standpoint.

3.3.1   MIG Logic Representation

Definition 3.1: An MIG is a homogeneous logic network with an indegree equal to 3 and each node representing the majority function. In a MIG, edges are marked by a regular or complemented attribute.

To determine some basic representation properties of MIGs, we compare them to the well-known *AND/OR/Inverter Graphs* (AOIGs) (which include AIGs). In terms of representation expressiveness, the elementary bricks in MIGs are *majority operators* while in AOIGs are conjunctions (AND) and disjunctions (OR). It is worth noticing that a majority operator $M(x, y, z)$ behaves as the conjunction operator $AND(x, y)$ when $z = 0$ and as the disjunction operator $OR(x, y)$ when $z = 1$. Therefore, majority is actually a generalization of both conjunction and disjunction. Recall that $M(x, y, z) = xy + xz + yz$. This property leads to the following theorem.

Theorem 3.1 *MIGs $\supset$ AOIGs.*

Proof In both AOIGs and MIGs, inverters are represented by complemented edge markers. An AOIG node is always a special case of a MIG node, with the third input biased to logic 0 or 1 to realize an AND or OR, respectively. On the other hand, a MIG node is never a special case of an AOIG node, because the functionality of the three input majority cannot be realized by a unique AND or OR.   ■

As a consequence of the previous theorem, MIGs are at least as good as AOIGs but potentially much better, in terms of representation compactness. Indeed, in the worst case, one can replace node-wise AND/ORs by majorities with the third input biased to a constant (0/1). However, even a more compact MIG representation can

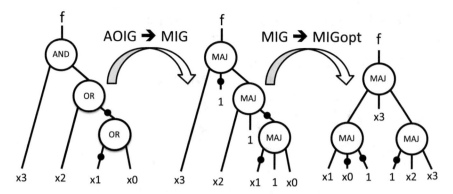

Fig. 3.2 MIG representation for $f = x_3 \cdot (x_2 + (x_1' + x_0)')$. Complementation is represented by bubbles on the edges

be obtained by fully exploiting its node functionality rather than fixing one input to a logic constant.

Figure 3.2 depicts a MIG representation example for $f = x_3 \cdot (x_2 + (x_1' + x_0)')$. The starting point is a traditional AOIG. Such AOIG has 3 nodes and 3 levels of depth, which is the best representation possible using just AND/ORs. The first MIG is obtained by a one-to-one replacement of AOIG nodes by MIG nodes. As shown by Fig. 3.2, a better MIG representation is possible by taking advantage of the majority function. This transformation will be detailed in the rest of this paper. In this way, one level of depth is saved with the same node count.

MIGs inherit from AOIGs some important properties, like universality and AIG inclusion. This is formalized by the following.

Corollary 3.1 *MIGs* $\supset$ *AIGs.*

Proof MIGs $\supset$ AOIGs $\supset$ AIGs $\implies$ MIGs $\supset$ AIGs.   ∎

Corollary 3.2 *MIG is an universal representation form.*

Proof MIGs $\supset$ AOIGs $\supset$ AIGs that are universal representation forms [8].   ∎

So far, we have shown that MIGs extend the representation capabilities of AOIGs. However, we need a proper set of manipulation tools to handle MIGs and automatically reach compact representations. For this purpose, we introduce hereafter a new Boolean algebra, based on MIG primitives.

3.3.2 MIG Boolean Algebra

We present a novel Boolean algebra, defined over the set $(\mathbb{B}, M, ', 0, 1)$, where M is the majority operator of three variables and $'$ is the complementation operator.

The following five primitive transformation rules, referred to as Ω, form an *axiomatic system* for $(\mathbb{B}, M,', 0, 1)$. All variables belong to $\mathbb{B}$.

$$\Omega \begin{cases} \textbf{Commutativity} : \Omega.C \\ M(x, y, z) = M(y, x, z) = M(z, y, x) \\ \textbf{Majority} : \Omega.M \\ \begin{cases} \text{if}(x = y) : M(x, x, z) = M(y, y, z) = x = y \\ \text{if}(x = y') : M(x, x', z) = z \end{cases} \\ \textbf{Associativity} : \Omega.A \\ M(x, u, M(y, u, z)) = M(z, u, M(y, u, x)) \\ \textbf{Distributivity} : \Omega.D \\ M(x, y, M(u, v, z)) = M(M(x, y, u), M(x, y, v), z) \\ \textbf{Inverter Propagation} : \Omega.I \\ M'(x, y, z) = M(x', y', z') \end{cases} \quad (3.2)$$

Axiom $\Omega.C$ defines a commutativity property. Axiom $\Omega.M$ declares a 2 over 3 decision threshold. Axiom $\Omega.A$ is an associative law extended to ternary operators. Axiom $\Omega.D$ establishes a distributive relation over majority operators. Axiom $\Omega.I$ expresses the interaction between M and complementation operators. It is worth noticing that $\Omega.I$ does not require operation type change like De Morgan laws, as it is well known from self-duality [1].

We prove that $(\mathbb{B}, M,', 0, 1)$ axiomatized by Ω is a Boolean algebra by showing that it induces a complemented distributive lattice [33].

Theorem 3.2 *The set* $(\mathbb{B}, M,', 0, 1)$ *subject to axioms in* Ω *is a Boolean algebra.*

Proof The system Ω embed median algebra axioms [34]. In such scheme, $M(0, x, 1) = x$ follows from $\Omega.M$. In [35], it is proved that a median algebra with elements 0 and 1 satisfying $M(0, x, 1) = x$ is a distributive lattice. Moreover, in our scenario, complementation is well defined and propagates through the operator M ($\Omega.I$). Combined with the previous property on distributivity, this makes our system a complemented distributive lattice. Every complemented distributive lattice is a Boolean algebra [33].∎

Note that there are other possible axiomatic systems for $(\mathbb{B}, M,', 0, 1)$. For example, one can show that in the presence of $\Omega.C$, $\Omega.A$ and $\Omega.M$, the rule in $\Omega.D$ is redundant [36]. In this work, we consider $\Omega.D$ as part of the axiomatic system for the sake of simplicity.

3.3.2.1 Derived Theorems

Several other complex rules, formally called theorems, in $(\mathbb{B}, M,', 0, 1)$ are derivable from Ω. Among the ones we encountered, three rules derived from Ω are of particular interest to logic optimization. We refer to them as Ψ and are described hereafter. In the following, the symbol $z_{x/y}$ represents a replacement operation, i.e., it replaces x with y in all its appearence in z.

$$\Psi \begin{cases} \textbf{Relevance} - \boldsymbol{\Psi}.R \\ M(x, y, z) = M(x, y, z_{x/y'}) \\ \textbf{Complementary Associativity} - \boldsymbol{\Psi}.C \\ M(x, u, M(y, u', z)) = M(x, u, M(y, x, z)) \\ \textbf{Substitution} - \boldsymbol{\Psi}.S \\ M(x, y, z) = \\ M(v, M(v', M_{v/u}(x, y, z), u), M(v', M_{v/u'}(x, y, z), u')) \end{cases}$$ (3.3)

The first rule, relevance ($\Psi.R$), replaces reconvergent variables with their neighbors. For example, consider the function $f = M(x, y, M(w, z', M(x, y, z)))$. Variables x and y are reconvergent because they appear in both the bottom and the top majority operators. In this case, relevance ($\Psi.R$) replaces x with y' in the bottom majority as $f = M(x, y, M(w, z', M(y', y, z)))$. This representation can be further reduced to $f = M(x, y, w)$ by using $\Omega.M$.

The second rule, complementary associativity ($\Psi.C$), deals with variables appearing in both polarities. Its rule of replacement is $M(x, u, M(y, u', z)) = M(x, u, M(y, x, z))$ as depicted by Eq. 3.3.

The third rule, substitution ($\Psi.S$), extends variable replacement to the non-reconvergent case. We refer the reader to Fig. 3.3 (appearing at page 75 of this dissertation) for an example about substitution ($\Psi.S$) applied to a 3-input parity function.

Hereafter, we show how Ψ rules can be derived from Ω.

Theorem 3.3 Ψ *rules are derivable by* Ω.

Proof **Relevance** ($\Psi.R$): Let S be the set of all possible input patterns for $M(x, y, z)$. Let $S_{x=y}$ ($S_{x=y'}$) be the subset of S such that $x = y$ ($x = y'$) condition is true. Note that $S_{x=y} \cap S_{x=y'} = \emptyset$ and $S_{x=y} \cup S_{x=y'} = S$. According to $\Omega.M$, variable z in $M(x, y, z)$ is only relevant for $S_{x=y'}$. Thus, it is possible to replace x with y', i.e., (x/y'), in all its appearance in z, preserving the original functionality.

Complementary Associativity ($\Psi.C$):
$M(x, u, M(u', y, z)) = M(M(x, u, u'), M(x, u, y), z)$ ($\Omega.D$)
$M(M(x, u, u'), M(x, u, y), z) = M(x, z, M(x, u, y))$ ($\Omega.M$)
$M(x, z, M(x, u, y)) = M(x, u, M(y, x, z))$ ($\Omega.A$)
Substitution ($\Psi.S$): We set $M(x, y, z) = k$ for brevity.
$k = M(v, v', k) = (\Omega.M)$
$M(M(u, u', v), v', k) = (\Omega.M)$
$M(M(v', k, u), M(v', k, u'), v) = (\Omega.D)$
Then, $M(v', k, u) = M(v', k_{v/u}, u)$ ($\Psi.R$)
and $M(v', k, u') = M(v', k_{v/u'}, u)$ ($\Psi.R$)
Recalling that $k = M(x, y, z)$, we finally obtain:
$M(x, y, z) = M(v, M(v', M_{v/u}(x, y, z), u), M(v', M_{v/u'}(x, y, z), u'))$. ∎

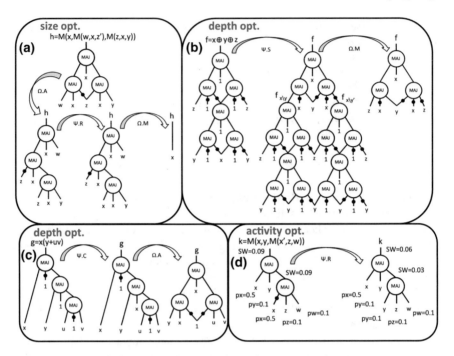

Fig. 3.3 Examples of MIG optimization for size, depth and switching activity

3.3.2.2 Soundness and Completness

The set $(\mathbb{B}, M,', 0, 1)$ together with axioms Ω and derivable theorems form our majority logic system. In a computer implementation of our majority logic system, the natural data structure for $(\mathbb{B}, M,', 0, 1)$ is a MIG and the associated manipulation tools are Ω and Ψ transformations. In order to be useful in practical applications, such as EDA, our majority logic system needs to satisfy fundamental mathematical properties such as soundness and completeness [32]. Soundness means that every argument provable by the axioms in the system is valid. This guarantees preserving of correctness. Completeness means that every valid argument has a proof in the system. This guarantees universal logic reachability. We show that our majority Boolean algebra is sound and complete.

Theorem 3.4 *The Boolean algebra* $(\mathbb{B}, M,', 0, 1)$ *axiomatized by* Ω *is sound and complete.*

Proof We first consider soundness. Here, we need to prove that all axioms in Ω are valid, i.e., preserve the true behavior (correctness) of a system. Rules $\Omega.C$ and $\Omega.M$ are valid because they express basic properties (commutativity and majority decision rule) of the majority operator. Rule $\Omega.I$ is valid because it derives from the self-duality of the majority operator. For rules $\Omega.D$ and $\Omega.A$, a simple way to prove their validity is to build the corresponding truth tables and check that they are actually the same. It is an easy exercise to verify that it is true. We consider now completeness. Here, we need to prove that every valid argument, i.e., $(\mathbb{B}, M, ', 0, 1)$-formula, has a proof in the system Ω. By contradiction, suppose that a true $(\mathbb{B}, M, ', 0, 1)$-formula, say α, cannot be proven true using Ω rules. Such $(\mathbb{B}, M, ', 0, 1)$-formula α can always be reduced by $\Psi.S$ rules into a $(\mathbb{B}, \cdot, +, ', 0, 1)$-formula. This is because $\Psi.S$ can behave as Shannon's expansion by setting $v = 1$ and u to a logic variable. Using Δ (Eq. 3.1), all $(\mathbb{B}, \cdot, +, ', 0, 1)$-formulas can be proven, including α. However, every $(\mathbb{B}, \cdot, +, ', 0, 1)$-formula is also contained by $(\mathbb{B}, M, ', 0, 1)$, where $\cdot$ and $+$ are emulated by majority operators. Moreover, rules in Ω with one input fixed to 0 and 1 behaves as Δ rules (Eq. 3.1). This means that also Ω is capable to prove the reduced $(\mathbb{B}, M, ', 0, 1)$-formula α, contradicting our assumption. Thus our system is sound and complete.   ∎

As a corollary of Ω soundness, all rules in Ψ are valid.

Corollary 3.3 Ψ *rules are valid in* $(\mathbb{B}, M, ', 0, 1)$.

Proof Ψ rules are derivable by Ω as shown in Theorem 3.3. Then, Ω rules are sound in $(\mathbb{B}, M, ', 0, 1)$ as shown in Theorem 3.4. Rules derivable from sound axioms are valid in the original domain.   ∎

As a corollary of Ω completeness, any element of a pair of equivalent $(\mathbb{B}, M, ', 0, 1)$-formulas, or MIGs, can be transformed one into the other by a sequence of Ω transformations. From now on, we use MIGs to refer to functions in the $(\mathbb{B}, M, ', 0, 1)$ domain. Still, the same arguments are valid for $(\mathbb{B}, M, ', 0, 1)$-formulas.

Corollary 3.4 *It is possible to transform any MIG α into any other logically equivalent MIG β, by a sequence of transformations in Ω.*

Proof MIGs are defined over the $(\mathbb{B}, M, ', 0, 1)$ domain. Following from Theorem 3.4, all valid arguments over $(\mathbb{B}, M, ', 0, 1)$ can be proved by a sequence of Ω rules. A valid argument is then $M(1, M(\alpha, \beta', 0), M(\alpha', \beta, 0)) = 0$ which reads "α is never different from β" according to the initial hypothesis. It follows that the sequence of Ω rules proving such argument is also logically transforming α into β.   ∎

3.3.2.3 Reachability

To measure the efficiency of a logic system, thus of its Boolean algebra, one can study (i) the ability to perform a desired task and (ii) the number of basic operations required to perform such a task. In the context of this work, the task we care about is logic optimization. For the graph size and graph depth metrics, MIGs can be smaller than AOIGs because of Theorem 3.1. However, the complexity of Ω sequences required to reach those desirable MIGs is not obvious. In this regard, we give an insight about the majority logic system efficiency by comparing the number of Ω rules needed to get an optimized MIGs with the number of Δ rules needed to get an evenly optimized AIGs. This type of efficiency metric is often referred to as reachability, i.e., the ability to reach a desired representation form with the smallest number of steps possible.

Theorem 3.5 *For a given optimization goal and an initial AOIG, the number of* Ω *rules needed to reach this goal with a MIG is smaller, or at most equal, than the number of* Δ *rules needed to reach the same goal with an AOIG.*

Proof Consider the shortest sequence of Δ rules meeting the optimization goal with an AOIG. On the MIG side, assume to start with the initial AOIG replacing node-wise AND/OR nodes with pre-configured majority nodes. Note that Ω rules with one input fixed to 0/1 behave as Δ rules. So, it is possible to emulate the same shortest sequence of Δ rules in AOIGs with Ω in MIGs. This is just an upper bound on the shortest sequence of Ω rules. Exploiting the full Ω expressiveness and MIG compactness, this sequence can be further shortened.   ∎

For a deeper theoretical study on majority logic expressiveness, we refer to [37]. In this work, we use the mathematical theory presented so far to define a consistent logic optimization framework. Then, we give experimental evidence on the benefits predicted by the theory. Results in Sect. 3.6 show indeed a depth reduction, over the state-of-the-art techniques, up to 48× thanks to our majority logic system. More details on the experiments are given in Sect. 3.6.

Operating on MIGs via the new Boolean algebra is one natural approach to run logic optimization. Interestingly enough, other approaches are also possible. In the following, we show how MIGs can be optimized exploiting other properties of the majority operator, such as bit-error masking.

3.3.3 Inserting Safe Errors in MIG

MIGs are hierarchical majority voting systems. One notable property of majority voting is the ability to correct different types of bit-errors. This feature is inherited by MIGs, where the error masking property can be exploited for logic optimization. The idea is to purposely introduce logic errors that are successively masked by the voting resilience in MIG nodes. If such errors are advantageous, in terms of logic simplifications, better MIG representations can be generated.

In the immediate following, we briefly review hereafter notations and definitions on logic errors [1, 38]. Then, we present the theoretical grounds for "safe error insertion" in MIGs. We define what type of errors, and at what overhead cost, can be introduced. Note that, in this work, we use the word *erroneous* to highlight the presence of a logic error. Our notation do not relate to testing or other fields.

Definition 3.1 The logic error in function f is defined as the difference between f and its erroneous version g and is computed as $f \oplus g$.

In principle, a logic error can be determined for any two circuits. In practical cases, a logic error is interpreted as a perturbation A on an original logic circuit f.

Notation A logic circuit f affected by error A is written f^A.

For example, consider the function $f = (a + b) \cdot c$. An error A defined as "*fix variable b to 0*" ($A: b = 0$) leads here to $f^A = ac$. In general, an error flips k entries in the truth table of the affected function. In the above example, $k = 1$.

To insert safe (permissible) errors in a MIG we consider a node w and we triplicate the sub-trees rooted at it. In each version of w we introduce logic errors heavily simplifying the MIG. Then, we use the three erroneous versions of w as inputs to a top majority node exploiting the error masking property. Unfortunately, a majority node cannot mask all types of errors. This limits our choice of permissible errors. *Orthogonal* errors, defined hereafter, fit our purposes. Informally, two logic errors are *orthogonal* if for any input pattern they cannot happen simultaneously. In the majority voting scenario the orthogonality is important because it guarantees that no two logic errors happen at the same time which would corrupt the original functionality.

Definition 3.2 Two logic errors A and B on a logic circuit f are said *orthogonal* if $(f^A \oplus f) \cdot (f^B \oplus f) = 0$.

To give an example of *orthogonal* errors consider again the function $f = (a+b) \cdot c$. Here, the two errors $A: a + b = 1$ and $B: c = 0$ are actually *orthogonal*. Indeed, by logic simplification, we get $(c \oplus f) \cdot (0 \oplus f) = (((a+b)c)'c + ((a+b)c)c') \cdot ((a+b)c) = ((a+b)c)'c \cdot ((a+b)c) = 0$. Instead, the errors $A: a + b = 1$ and $B: c = 1$ are not *orthogonal* for f. This is because the input $(1, 1, 1)$ triggers both errors.

Now consider back a generic MIG root w. Let A, B and C be three pairwise *orthogonal* errors on w. Being all pairwise *orthogonal*, a top majority node $M(w^A, w^B, w^C)$ is capable to mask A, B and C orthogonal errors restoring the original functionality of w. This is formalized in the following theorem.

Theorem 3.6 *Let w be a generic node in a MIG. Let A, B and C be three pairwise orthogonal errors on w. Then the following equation holds: $w = M(w^A, w^B, w^C)$*

Proof The equation $w = M(w^A, w^B, w^C)$ is logically equivalent to $w \oplus M(w^A, w^B, w^C) = 0$. The $\oplus$ (XOR) operator propagates into the majority operator as $w \oplus M(w^A, w^B, w^C) = M(w^A \oplus w, w^B \oplus w, w^C \oplus w)$. Recalling that $M(a, b, c) = ab + ac + bc$ we rewrite the previous expression as $(w^A \oplus w) \cdot (w^B \oplus w) + (w^A \oplus w) \cdot (w^C \oplus w) + (w^B \oplus w) \cdot (w^C \oplus w)$. Recall the previously introduced definition of *orthogonal* errors $(f^A \oplus f) \cdot (f^B \oplus f) = 0$. As all errors here are pairwise *orthogonal*, we have that each term $(w^{err_1} \oplus w) \cdot (w^{err_2} \oplus w)$ is 0 because of the aforementioned definition, so $0 + 0 + 0 = 0$. Thus, $w \oplus M(w^A, w^B, w^C) = 0$. ∎

Note that a MIG $w = M(w^A, w^B, w^C)$ can have up to three times the size and one more level of depth as the original w. This means that simplifications enabled by *orthogonal* errors A, B and C must be significant enough to compensate for such overhead. Note also that this approach resembles triple modular redundancy [39] and its approximate variants [40], but operates differently. Here, we exploit the error masking property in majority operators to enable logic simplifications rather than covering potential hardware failures. More details on how to identify advantageous *orthogonal* errors in MIGs will be given in Sect. 3.5.1 together with related Boolean optimization methods.

In the following sections, we present algorithms for algebraic and Boolean optimization of MIGs.

3.4 MIG Algebraic Optimization

In this section, we propose algebraic optimization methods for MIGs. They exploit axioms and derived theorems of the novel Boolean algebra. Our algebraic optimization procedures target size, depth and switching activity reduction in MIGs.

3.4.1 Size-Oriented MIG Algebraic Optimization

To optimize the size of a MIG, we aim at reducing the number of its nodes. Node reduction can be done, at first instance, by applying the majority rule. In the MIG Boolean algebra domain this corresponds to the evaluation of the majority axiom $(\Omega.M)$ from *Left to Right* $(L \rightarrow R)$, as $M(x, x, z) = x$. A different node elimination opportunity arises from the distributivity axiom $(\Omega.D)$, evaluated from *Right to Left* $(R \rightarrow L)$, as $M(x, y, M(u, v, z)) = M(M(x, y, u), M(x, y, v), z)$. By applying $\Omega.M_{L \rightarrow R}$ and $\Omega.D_{R \rightarrow L}$ to all MIG nodes, in an arbitrary sequence, we can actually eliminate nodes. By repeating this procedure until no improvement exists, we designed a simple yet powerful procedure to reduce a MIG size. Embedding some intelligence in the graph exploration direction, e.g., the sequence of MIG nodes, immediately improves the optimization effectiveness. Note that the applicability of majority and distributivity depends on the particular MIG structure. Indeed, there may be MIGs where no direct node elimination is evident. This is because (i) the optimal size is reached or (ii) we are stuck in a local minimum. In the latter case, we want to reshape the MIG in order to encode new reduction opportunities. The rationale driving the reshaping process is to locally increase the number of common inputs/variables to MIG nodes. For this purpose, the associativity axioms $(\Omega.A, \Psi.C)$ allow us to move variables between adjacent levels and the relevance axiom $(\Psi.R)$ to exchange reconvergent variables. When a more radical transformation is beneficial, the substitution axiom $(\Psi.S)$ replaces pairs of independent variables, temporarily inflating the MIG. Once the reshaping process has created new reduction opportunities, majority $(\Omega.M_{L \rightarrow R})$ and distributivity $(\Omega.D_{R \rightarrow L})$ are applied again over the MIG to simplify it. The reshaping and elimination processes can be iterated over a user-defined number of cycles, called *effort*. Such MIG-size algebraic optimization strategy is summarized in Algorithm 2.

Algorithm 2 MIG Algebraic Size-Optimization Pseudocode

INPUT: MIG α **OUTPUT:** Optimized MIG α.

  for (cycles=0; cycles<*effort*; cycles++) **do**

    $\Omega.M_{L \to R}(\alpha)$; $\Omega.D_{R \to L}(\alpha)$;

    $\Omega.A(\alpha)$; $\Psi.C(\alpha)$;

    $\Psi.R(\alpha)$; $\Psi.S(\alpha)$; } reshape } eliminate

    $\Omega.M_{L \to R}(\alpha)$; $\Omega.D_{R \to L}(\alpha)$;

  end for

For the sake of clarity, we comment on the MIG-size algebraic optimization of a simple example, reported in Fig. 3.3a. The input MIG is equivalent to the formula $M(x, M(x, z', w), M(x, y, z))$, which has no evident simplification by majority and distributivity axioms. Consequently, the reshape process is invoked to locally increase the number of common inputs. Associativity $\Omega.A$ swaps w and $M(x, y, z)$ in the original formula obtaining $M(x, M(x, z', M(x, y, z)), w)$, when variables x and z are close to the each other. After that, the relevance $\Psi.R$ modifies the inner formula $M(x, z', M(x, y, z))$, exchanging variable z with x and obtaining $M(x, M(x, z', M(x, y, x)), w)$. At this point, the final elimination process is applied, simplifying the reshaped representation as $M(x, M(x, z', M(x, y, x)), w) = M(x, M(x, z', x), w) = M(x, x, w) = x$ by using $\Omega.M_{L \to R}$.

3.4.2 Depth-Oriented MIG Algebraic Optimization

To optimize the depth of a MIG, we aim at reducing the length of its critical path. A valid strategy for this purpose is to move late arrival (critical) variables close to the outputs. In order to explain how critical variables can be moved, while preserving the original functionality, consider the general case in which a part of the critical path appears in the form $M(x, y, M(u, v, z))$. If the critical variable is x, or y, no simple move can reduce the depth of $M(x, y, M(u, v, z))$. Whereas, if the critical variable belongs to $M(u, v, z)$, say z, depth reduction is achievable. We focus on the latter case, with order $t_z > t_u \geq t_v > t_x \geq t_y$ for the variables arrival time (depth). Such an order can arise from (i) an unbalanced MIG whose inputs have equal arrival times, or (ii) a balanced MIG whose inputs have different arrival times. In both cases, z is the critical variable arriving later than u, v, x, y, hence the local depth is $t_z + 2$. If we apply the distributivity axiom $\Omega.D$ from left to right ($L \to R$), we obtain $M(x, y, M(u, v, z)) = M(M(x, y, u), M(x, y, v), z)$ where z is pushed one level up, reducing the local depth to $t_z + 1$. Such technique is applicable to a broad range of cases, as all the variables appearing in $M(x, y, M(u, v, z))$ are distinct and independent. However, there is a size penalty of one extra node. In the favorable cases for which associativity axioms ($\Omega.A, \Psi.C$) apply, critical variables can be pushed up with no penalty. Furthermore, where majority axiom applies $\Omega.M_{L \to R}$, it is possible to reduce both depth and size. As noted earlier, there exist cases for which moving critical variables cannot improve the overall depth. This is because (i) the optimal depth is reached or (ii) we are stuck in a local minimum. To move away from a local minimum, the reshape process is useful. The reshape and critical variable push-up processes can be iterated over a user-defined number of cycles, called *effort*. Such MIG-depth algebraic optimization strategy is summarized in Algorithm 3.

Algorithm 3 MIG Algebraic Depth-Optimization Pseudocode

INPUT: MIG α **OUTPUT:** Optimized MIG α.

  for (cycles=0; cycles<*effort*; cycles++) **do**
    $\Omega.M_{L \to R}(\alpha); \Omega.D_{L \to R}(\alpha); \Omega.A(\alpha);$ ⎫
    $\Omega.A(\alpha); \Psi.C(\alpha);$
    $\Psi.R(\alpha); \Psi.S(\alpha);$ ⎬ reshape ⎫ push-up
    $\Omega.M_{L \to R}(\alpha); \Omega.D_{L \to R}(\alpha); \Omega.A(\alpha);$ ⎭ ⎭
  end for

We comment on the MIG-depth algebraic optimization using two examples depicted by Fig. 3.3b, c. The considered functions are $f = x \oplus y \oplus z$ and $g = x(y + u \cdot v)$ with initial MIG representations derived from their optimal AOIGs. In both of them, all inputs have 0 arrival time. No direct push-up operation is advantageous. The reshape process is invoked to move away from local minimum. For $g = x(y + uv)$, complementary associativity $\Psi.C$ enforces variable x to appear in two adjacent levels, while for $f = x \oplus y \oplus z$ substitution $\Psi.S$ replaces x with y, temporarily inflating the MIG. After this reshaping, the push-up procedure is applicable. For $g = x(y + u \cdot v)$, associativity $\Omega.A$ exchanges $1'$ with $M(u, 1', v)$ in the top node, reducing by one level the MIG depth. For $f = x \oplus y \oplus z$, majority $\Omega.M_{L \to R}$ heavily simplifies the structure and reduces the intermediate MIG depth by four levels. The optimized MIGs are much shorter than their optimal AOIGs counterparts. Note that Algorithm 3 produces irredundant solutions.

3.4.3   Switching Activity-Oriented MIG Algebraic Optimization

To optimize the total switching activity of a MIG, we aim at reducing (i) its size and (ii) the probability for nodes to switch from logic 0 to 1, or vice versa. For the size reduction task, we can run the same MIG-size algebraic optimization described previously. To minimize the switching probability, we want that nodes do not change values often, i.e., the probability of a node to be logic 1 (p_1) is close to 0 or 1 [41]. For this purpose, relevance $\Psi.R$ and substitution $\Psi.S$ can exchange variables with undesirable $p_1 \sim 0.5$ with more favorable variables having $p_1 \sim 1$ or $p_1 \sim 0$. In Fig. 3.3d, we show an example where relevance $\Psi.R$ replaces a variable x having $p_1 = 0.5$ with a reconvergent variable y having $p_1 = 0.1$, thus reducing the overall MIG switching activity.

3.5 MIG Boolean Optimization

In this section, we propose Boolean optimization methods for MIGs. They exploit the *safe error insertion schemes* presented in Sect. 3.3.3. First, we introduce two techniques to identify advantageous orthogonal errors in MIGs. Second, we present our Boolean optimization technique targeting depth and size reduction in MIGs. Note that also other optimization goals are possible.

3.5.1 Identifying Advantageous Orthogonal Errors in MIGs

In the following, we present two methods for identifying advantageous triplets of *orthogonal* errors in MIGs.

3.5.1.1 Critical Voters Method

A natural way to discover advantageous triplets of *orthogonal* errors is to analyze a MIG structure. We want to identify critical portions of a MIG to be simplified thanks to these errors. To do so, we focus on nodes[1] that have the highest impact on the final voting decision, i.e., influencing a Boolean function most. We call such nodes *critical voters* of a MIG. Critical voters can also be primary input themselves. To determine the critical voters, we rank MIG nodes based on a *criticality* metric. The *criticality* computation goes as follows. Consider a MIG node m. We label all MIG nodes whose computation depends on m. For all such nodes, we calculate the impact of m by propagating a unit weight value from m outputs up to the root with an attenuation factor of $1/3$ each time a majority node is encountered. We finally sum up all the values obtained and call this result *criticality* of m. Intuitively, MIG nodes with the highest *criticality* are critical voters.

For the sake of clarity, we give an example of *criticality* computation in Fig. 3.4. Node $m5$ has *criticality* of 0, since it is the root and does not propagate to any node. Node $m4$ has *criticality* of $1/3$ (a unit weight propagated to $m5$ and attenuated by $1/3$). Node $m3$ has *criticality* of $1/3$ ($m4$) + $(1/3+1)/3$ (direct and $m4$ contribution to $m5$) which sums up to $7/9$. Node $m2$ has *criticality* of $1/3$ ($m3$) + $4/9$ ($m4$) + $7/27$ ($m5$) which sums up to $28/27$. Node $m1$ has *criticality* $1/3$ + *criticality* of $m2$ attenuated by factor 3 which sums up to about $2/3$. Among the inputs, only $x1$ has a notable *criticality* being $1/3$ ($m3$) + $1/9$ ($m4$) + $(1/3+1/9+1)/3$ ($m5$) which sums up to $25/27$. Here the two elements with highest *criticality* are $m2$ and $x1$.

We first determine two critical voters a and b and a set of MIG nodes fed directly by both a and b, say $\{c_1, c_2, \ldots, c_n\}$. In this context, an advantageous triplet of *orthogonal* errors is: A: $a = b'$, B: $c_1 = a, c_2 = a, \ldots, c_n = a$ and C: $c_1 = b, c_2 =$

---

[1]In the context of the critical voters technique we consider also the primary inputs to be a special class of nodes with no fan-in.

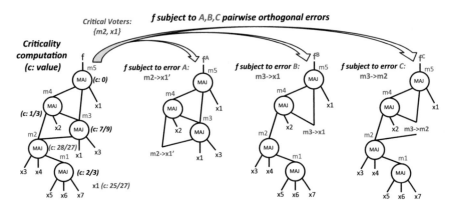

Fig. 3.4 Example of *criticality* computation and *orthogonal* errors

$b, \ldots, c_n = b$. Consider again the example in Fig. 3.4. There, the critical voters are $a = m2$ and $b = x1$, while $c_1 = m3$. Thus, the pairwise *orthogonal* errors are $m2 = x1'$ (A), $m3 = x1$ (B) and $m3 = m2$ (C), as shown in Fig. 3.4. The actual *orthogonality* of A, B and C type of errors is proved in the following theorem.

Theorem 3.7 *Let a and b be two critical voters in a MIG. Let* $\{c_1, c_2, \ldots, c_n\}$ *be the set of MIG nodes fed by both a and b in the same polarity. Then, the following errors are pairwise* orthogonal: *A:* $a = b'$, *B:* $c_1 = a, c_2 = a, \ldots, c_n = a$ *and C:* $c_1 = b, c_2 = b, \ldots, c_n = b$.

Proof Starting from a MIG w, we build the three erroneous versions w^A, w^B and w^C as described above. We show that *orthogonality* holds for all 3 pairs. **Pair** $(\boldsymbol{w}^A, \boldsymbol{w}^B)$: We need to show that $(w^A \oplus w) \cdot (w^B \oplus w) = 0$. The element $w^A \oplus w$ implies $a = b$, being the difference between the original and the erroneous one with $a = b'$ ($a \neq b$). The element $w^B \oplus w$ implies $c_i \neq a$ ($c_i = a'$), being the difference between the original and the erroneous one with $c_i = a$. However, if $a = b$ then c_i cannot be a' because $c_i = M(a, b, x) = M(a, a, x) = a \neq a'$ by construction. Thus, the two elements cannot be true at the same time, making $(w^A \oplus w) \cdot (w^B \oplus w) = 0$. **Pair** $(\boldsymbol{w}^A, \boldsymbol{w}^C)$: This case is analogous to the previous one. **Pair** $(\boldsymbol{w}^B, \boldsymbol{w}^C)$: We need to show that $(w^B \oplus w) \cdot (w^C \oplus w) = 0$. As we deduced before, the element $w^B \oplus w$ implies $c_i \neq a$ ($c_i = a'$). Similarly, the element $w^C \oplus w$ implies $c_i \neq b$ ($c_i = b'$). By the *transitive property of equality and congruence* in the Boolean domain $c_i \neq a$ and $c_i \neq b$ implies $a = b$. However, if $a = b$, then $c_i = M(a, b, x) = M(a, a, x) = M(b, b, x) = a = b$ which contradicts both $c_i \neq a$ and $c_i \neq b$. Thus, w^B, w^C cannot be true simultaneously, making $(w^B \oplus w) \cdot (w^C \oplus w) = 0$. ∎

Even though focusing on critical voters is typically a good strategy for safe error insertion in MIGs, sometimes other techniques can be also convenient. In the following, we present one of these alternative techniques.

3.5.1.2 Input Partitioning Method

As a complement to critical voters method, we propose a different way to derive advantageous triplets of *orthogonal* errors in MIGs. In this case, we focus on the inputs rather than looking for internal MIG nodes. In particular, we search for inputs leading to advantageous simplifications when erroneous. Analogously to the *criticality* metric in critical voters, we use here a decision metric, called *dictatorship* [42], to select the most profitable inputs for logic error insertion. The *dictatorship* is the ratio of input patterns over the total (2^n) for which the output assumes the same value than the selected input [42]. For example, in the function $f = (a + b) \cdot c$, the inputs a and b have equal *dictatorship* of 5/8 while input c has an higher *dictatorship* of 7/8. The inputs with the highest *dictatorship* are the ones where we want to insert logic errors. Indeed, they have the largest influence on the circuit functionality and its structure.

Exact computation of the *dictatorship* requires exhaustive simulation of an MIG structure, which is not feasible for practical reasons. Heuristic approaches to estimate *dictatorship* involve partial random simulation and graph techniques [42].

After exact or heuristic computation of the dictatorship, we select a subset of the primary inputs with highest *dictatorship*. Next, for each selected input, we determine a condition that causes an error. We require these errors to be *orthogonal*. Since we operate directly on the primary inputs, we already divide the Boolean space into disjoint subsets that are *orthogonal*. Because we need at least three errors, we need to consider at least three inputs to be made erroneous, say x, y and z. A possible partition is the following: $\{x \neq y, x = y = z, x = y = z'\}$. The corresponding errors are A: $x = y$ for $\{x \neq y\}$, B: $z = y'$ when $x = y$ for $\{x = y = z\}$ and C: $z = y$ when $x = y$ for $\{x = y = z'\}$. We formally prove A, B and C orthogonality hereafter.

Theorem 3.8 *Consider the input split* $\{x \neq y, x = y = z, x = y = z'\}$ *in a MIG. Three errors* A, B *and* C *selectively affecting one subset but not the others are pairwise* orthogonal.

Proof To prove the theorem it is sufficient to show that the split $\{x \neq y, x = y = z, x = y = z'\}$ is actually a partition of the whole Boolean space, i.e., a union of disjoint (non-overlapping) subsets. It is an easy exercise to enumerate all the eight possible $\{x, y, z\}$ input patterns and associate with each of them the corresponding $\{x \neq y, x = y = z, x = y = z'\}$ subset. By doing so, one can see that no $\{x, y, z\}$ pattern is associated with more than one sub-set, meaning that all subsets are disjoint. Moreover, all together, they form the whole Boolean space.   ∎

For the sake of clarity, we report an illustrative example on the input partitioning method. The function is $f = M(x, M(x, y', z), M(x', y, z))$. The input split is $\{x \neq y, x = y = z, x = y = z'\}$ which is affected by errors A, B and C, respectively. The first error A imposes $x = y$ leading to $f^A = M(x, M(y, y', z), M(x', x, z))$ which can be further simplified into $f^A = M(x, z, z) = z$ by $\Omega.M$. The second error B imposes $z = y'$ when $x = y$. This is the case for the bottom level majority operators

$M(x, y', z)$ and $M(x', y, z)$ which are transparent when $x = y$. Therefore, error B leads to $f^B = M(x, M(x, y', y'), M(x', y, y'))$ which can be further simplified into $f^B = M(x, y', x') = y'$ by $\Omega.M$. The third error C imposes $z = y$ when $x = y$ holds. Analogously to error B, error C leads to $f^C = M(x, M(x, y', y), M(x', y, y))$ which can be further simplified into $f^C = M(x, x, y) = x$ by $\Omega.M$. A top majority node finally merges the three functions into $f = M(f^A, f^B, f^C) = M(z, y', x)$ which correctly represents the objective function but has 2 fewer nodes and 1 level less than the original representation.

3.5.2 Depth-Oriented MIG Boolean Optimization

The most intuitive way to exploit safe error insertion in MIGs is to reduce the number of levels. This is because the initial overhead in $w = M(w^A, w^B, w^C)$, where w is the initial MIG and w^A, w^B, w^C are the three erroneous versions, is just one additional level. This extra level is usually amply recovered during simplification and optimization of MIG erroneous branches. For depth-optimization purposes, the critical voters method introduced in Sect. 3.3.3 enables very good results. The reason is the following. Critical voters appear along the critical path more than once. Thus, the possibility to insert simplifying errors on critical voters directly enables a strong reduction in the maximum number of levels. Sometimes, using an actual MIG root for error insertion requires an unpractical size overhead. In these cases, we bound the critical voters search to sub-MIGs partitioned on a depth criticality basis. Once the critical voters and a proper error insertion root have been identified, three erroneous sub-MIG versions are generated as explained in Sect. 3.3.3. On these sub-MIGs, we want to reduce the logic height. We do so by running algebraic MIG optimization on them (Algorithm 3). Note that, in principle, also MIG Boolean methods can be re-used. This would correspond to a recursive Boolean optimization. However, it turned out during experimentation that algebraic optimizations already produce satisfactorily results at the local level. Thus, it makes more sense to apply Boolean techniques iteratively on the whole MIG structure rather than recursively on the same logic portion. At the end of the optimization of erroneous branches, the new MIG-roots must be given in input to a top majority voting node. This re-establishes the functional correctness. A *last gasp* of MIG algebraic optimization is applied at this point, to take advantage of the simplification opportunities arosen from the integration of erroneous branches. This Boolean optimization strategy is summarized in Algorithm 4.

We comment on the MIG Boolean depth optimization with a simple example, reported in Fig. 3.5. First, the critical voters are searched and identified, being in this example the input $x1$ and the node $m2$ (from Fig. 3.4). The proper error insertion root in this small example is the MIG root itself. So, three different versions of the root f are generated with errors $f^{m2/x1'}$, $f^{m3/m2}$ and $f^{m3/x1}$. Each erroneous branch is handled by fast algebraic optimization to reduce its depth. The detailed algebraic optimization steps involved are shown in Fig. 3.5. The most common operation is

Algorithm 4 MIG Boolean Depth-Optimization Pseudocode

INPUT: MIG α **OUTPUT:** Optimized MIG α.

for (cycles=0; cycles<*effort*; cycles++) **do**

$\{a, b\}$=search_critical_voters(α);// Critical voters a, b searched

c=size_bounded_root(α, a, b);// Proper error insertion root

x_1^n=common_parents(α, a, b);// Nodes fed by both a and b

c^A=$c^{b/a'}$;// First erroneous branch

c^B=$c^{x_1^n/a}$;// Second erroneous branch

c^C=$c^{x_1^n/b}$;// Third erroneous branch

MIG-depth_Alg_Opt(c^A);// Reduce the erroneous branch height

MIG-depth_Alg_Opt(c^B);// Reduce the erroneous branch height

MIG-depth_Alg_Opt(c^C);// Reduce the erroneous branch height

c=$M(c^A, c^B, c^C)$;// Link the erroneous branches

MIG-depth_Alg_Opt(c); // Last Gasp

if depth(c) is not reduced **then**

revert to previous MIG state;

end if

end for

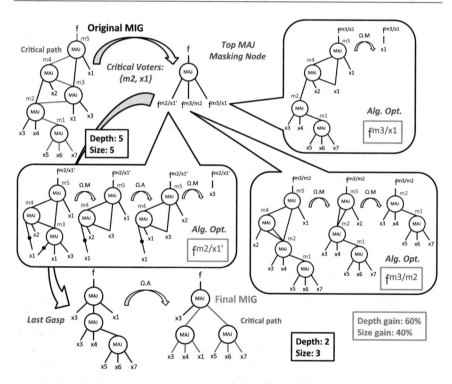

Fig. 3.5 MIG boolean depth-optimization example based on critical voters errors insertion. Final depth reduction: 60%

$\Omega.M$ that directly simplifies the introduced errors. The optimized erroneous branches are then linked together by a top fault-masking majority node. A last gasp of algebraic optimization on the final MIG structure further optimizes its depth. In summary, our MIG Boolean optimization techniques attains a depth reduction of 60 %.

3.5.3 Size-Oriented MIG Boolean Optimization

Safe error insertion in MIGs can be used for size reduction. In this case, the branch triplication overhead in $w = M(w^A, w^B, w^C)$ imposes tight simplification requirements. One way to handle this situation is to enforce stricter selection metrics on critical voters. However, the benefits deriving from this approach are limited. A better solution is to change the type of error inserted and use the *input partitioning method*. Indeed, the *input partitioning method* can focus on the most influent inputs of a MIG, and introduces selective simplification on them. The resulting Boolean optimization procedure is similar to Algorithm 3 but with depth techniques replaced by size techniques, and critical voter search replaced by input partitioning methods.

3.6 Experimental Results

In this section, we test the performance of our MIG optimization techniques on academic and industrial benchmarks. We run logic optimization experiments (comparing logic networks) and complete design experiments (consisting of logic synthesis and physical design) on commercial ASIC and FPGA flows.

3.6.1 Methodology

We developed a majority-logic manipulation package, called *MIGhty*, consisting of about 8 k lines of C code. It embeds various optimization commands based on the theory presented so far. In this work, we use a particular *MIGhty* optimization strategy targeting strong depth reduction interleaved with size recovery phases. The top-level optimization script is depicted by Algorithm 5. This technique starts by reducing the depth by algebraic methods implying a small size overhead. After a fast reshaping step, it decreases the size of the MIG by level-bounded size reduction. At this point, Boolean MIG depth optimization is invoked to significantly reduce the number of levels at the price of a temporary MIG size inflation. Further level reduction opportunities are exploited in an algebraic depth reduction step. Then, size recovery is achieved by Boolean intertwined with algebraic size reduction. A small depth overhead is possible in this phase due to the size reduction. Finally, a last gasp of algebraic depth optimization further compacts the MIG followed by level-bounded algebraic size reduction. All optimization steps have a runtime complexity linear w.r.t. the MIG size, i.e., are imposed to consider each node at least once.

The script in Algorithm 5 is a composite optimization strategy, similarly to the class of *resyn* scripts in ABC [27].

Algorithm 5 Top-Level MIG-optimization Script

INPUT: MIG α.   **OUTPUT:** Optimized MIG α.

MIG-depth_Alg_Opt(α);// small size overhead
MIG-reshaping(α);// reshuffling
MIG-size_Alg_Opt(α);// no depth overhead
MIG-depth_Bool_Opt(α);// pronounced size overhead
MIG-reshaping(α);// reshuffling
MIG-depth_Alg_Opt(α);// small size overhead
MIG-size_Bool_Opt(α);// small depth overhead
MIG-size_Alg_Opt(α);// no depth overhead
MIG-reshaping(α);// reshuffling
MIG-depth_Alg_Opt(α);// small size overhead
MIG-size_Alg_Opt(α);// no depth overhead

MIGhty reads files in Verilog or AIGER format and writes back a Verilog description of the optimized MIG. In order to simplify successive mapping steps, *MIGhty* reduces majority functions into AND/ORs if no size/depth overhead is implied. Thus, only the essential majority functions are written. Also, the number of inversions is minimized by $\Omega.I$ before writing.

We consider IWLS'05 Open Cores benchmarks and larger arithmetic HDL benchmarks. As a case study, we also consider various adder circuits. All the Verilog files deriving from our experiments can be downloaded at [43], for the sake of reproducibility. In all benchmarks, we assumed the input signals to be available at time 0. In total, we optimized about half a million equivalent gates over 31 benchmarks.

For the pure logic optimization experiments, we use as reference tool the ABC academic synthesizer [27], with the delay oriented script $if - g$; $iresyn$. The initial $if - g$ optimization strongly reduces the AIG depth by using SOP-balancing [44]. The latter $iresyn$ optimization performs fast rewriting passes on the AIG, reducing mostly the number of nodes but potentially also the number of levels.

We chose the AIG script $if - g$; $iresyn$ because its optimization rationale is close to our MIG optimization strategy and the respective runtimes are comparable. Note that ABC offers many other optimization scripts. Some of them may give better results under determinate conditions (benchmark type, size etc.). As the purpose of this work is primarily to assess the potential of MIG optimization w.r.t. to analogous AIG optimization, we neglect considerations and comparisons related to other ABC commands.

While comparing size and depth of MIGs versus AIGs already gives some good intuition on a data structure and optimization effectiveness, we aim at providing results on even grounds. For this reason, we map both AIG-optimized and MIG-optimized circuits onto LUT6. We perform LUT mapping using the established ABC script $dch - f; if -m -K$ 6.

For the complete design experiments, we consider a 22-nm (28-nm) commercial ASIC (FPGA) flow suite. The commercial flow consists of a logic synthesis step followed by place and route. In this case, we use the MIG-optimized Verilog file as input to the commercial tools in place of the original Verilog file. In other words, the *MIGhty* package operates as a front-end to the flows. Indeed, the efficiency of MIG-optimization helps the commercial tool to design better circuits. With the final circuit speed being our main design goal, we use an *ultra-high delay effort script* in the commercial tools.

3.6.2 Optimization Case Study: Adders

We first test the MIG optimization capabilities for adders, that are known hard-to-optimize circuits [45].

Results for more general benchmarks are given in the next subsection. Table 3.1 shows the adder results. Our optimized MIG adders have 4 to 48× smaller depth than the original AIGs. In all cases, the optimized MIG structure achieves depths close to the ones of carry-look ahead adders. Considering LUT mapped results, MIG-optimization enables significantly less deep circuits, having 1.75 to 14× smaller depth than LUT6 circuits mapped from the original AIGs.

3.6.3 General Optimization Results

Table 3.2 shows general results for *MIGhty* logic optimization and LUT-6 mapping. For the IWLS'05 and HDL arithmetic benchmarks, we see a total improvement in all size, depth and switching activity metrics, w.r.t. to AIG optimized by ABC. The switching activity is computed by the ABC command *ps -p*. The same improvement trend holds also for LUT mapped circuits. Since logic depth was our main optimization target, we notice there the largest reduction.

Table 3.1 Adder optimization results

Type	Bit	Orig. AIG		Map. AIG		Opt. MIG		Map. MIG	
		size	lvl	lut6	lvl	size	lvl	lut6	lvl
2-op	32	352	96	65	13	610	12	150	4
2-op	64	704	192	132	26	1159	11	272	5
2-op	128	1408	384	267	52	14672	19	3684	7
2-op	256	2816	768	544	103	7650	16	1870	7
3-op	32	760	68	127	14	1938	16	349	8
4-op	64	1336	136	391	27	2212	18	524	7

Table 3.2 MIG Logic Optimization and LUT-6 Mapping Results

Benchmark	I/O	MIGhty					ABC				
		Opt. MIG		Map. MIG		Runtime	Opt. AIG		Map. AIG		Runtime
		Size	Depth	LUT6	Depth	(s)	Size	Depth	LUT6	Depth	(s)
Open Cores IWLS'05											
DSP	4223/3953	40517	34	11077	11	7.98	39958	41	11309	12	5.39
ac97_ctrl	2255/2250	10745	8	2917	3	6.52	10497	9	2914	3	8.98
aes_core	789/668	20947	18	3902	4	11.78	20632	19	3754	5	8.22
des_area	368/72	4186	22	735	6	1.04	5043	24	1012	7	2.11
des_perf	9042/9038	67194	15	12796	3	34.22	75561	15	12814	3	25.43
ethernet	10672/10696	57959	15	18108	6	23.69	56882	22	18267	6	36.54
i2c	147/142	971	8	270	3	0.11	1009	10	268	4	0.05
mem_ctrl	1198/1225	7143	19	2333	7	0.38	9351	22	2582	7	0.33
pci_bridge32	3519/3528	18063	16	5294	6	3.28	16812	18	5424	7	2.22
pci_spoci_ctrl	85/76	932	11	276	4	0.04	994	13	287	4	0.02
sasc	133/132	621	6	152	2	0.11	657	7	152	2	0.03
simple_spi	148/147	837	8	206	3	0.05	770	10	211	3	0.01
spi	274/276	3337	19	812	6	3.71	3430	24	854	7	2.28
ss_pcm	106/98	397	6	104	2	0.01	381	6	104	2	0.01
systemcaes	930/819	9547	25	1845	7	5.26	11014	31	2060	8	4.79
systemcdes	314/258	2453	19	515	5	2.21	2495	21	623	5	1.05
tv80	373/404	7397	30	1980	11	6.43	7838	35	2036	11	2.97
usb_funct	1860/1846	12995	19	3333	5	13.45	13914	20	3394	5	9.04

(continued)

Table 3.2 (continued)

Benchmark	I/O	MIGhty					ABC				
		Opt. MIG		Map. MIG		Runtime (s)	Opt. AIG		Map. AIG		Runtime (s)
		Size	Depth	LUT6	Depth		Size	Depth	LUT6	Depth	
Open Cores IWLS'05											
usb_phy	113/111	372	7	136	2	0.11	380	7	136	2	0.05
IWLS'05 total		266613	**305**	66791	**96**	120.38	277618	**354**	68201	**103**	109.52
Arithmetic HDL		Size	Depth	LUT6	Depth	Runtime (s)	Size	Depth	LUT6	Depth	Runtime (s)
MUL32	64/64	9096	36	1852	10	2.90	8903	40	1993	11	1.90
sqrt32	32/16	2171	164	544	54	1.02	1353	292	236	55	1.22
diffeq1	354/289	17281	219	4685	45	56.32	21980	235	4939	45	16.88
div16	32/32	4374	102	818	37	4.67	5111	132	806	38	2.44
hamming	200/7	2071	61	517	14	2.01	2607	73	590	17	2.54
MAC32	96/65	9326	41	2095	11	4.30	9099	54	2044	12	7.76
metric_comp	279/193	18493	77	6202	29	16.21	21112	95	6796	31	9.51
revx	20/25	7516	143	1937	40	10.70	7516	162	2176	42	12.02
mul64	128/128	25773	109	6557	31	13.84	26024	186	6751	43	10.09
max	512/130	4210	29	1023	12	1.67	2964	113	818	20	2.23
square	64/127	17550	40	4393	13	18.66	17066	168	4278	35	12.24
log2	32/32	31326	201	8809	59	23.32	30701	272	8223	73	15.54
Arithmetic total		149727	**1222**	39432	**355**	155.62	154436	**1822**	39650	**422**	94.37

Considering the IWLS'05 benchmarks, that are large but not deep, *MIGhty* enables about 14 % depth reduction. At the LUT-level, we see about 7 % depth reduction. At the same time, the size and switching activity are reduced by about 4 and 2 %, respectively. At the LUT-level, size and switching activity are reduced by about 2 and 1 %, respectively.

Focusing on the arithmetic HDL benchmarks, we see a better depth reduction. Here, *MIGhty* enables about 33 % depth reduction. At the LUT-level, it enables about 16 % depth reduction. At the same time, *MIGhty* reduces size and switching activity by 4 and 0.1 %. At the LUT-level, this corresponds to about 1 % size reduction and practically the same switching activity.

The switching activity numbers are not reported in Table 3.2 for space reasons but can be reproduced using the ABC command *ps -p* on the files downloadable at [43].

Table 3.2 confirms that the runtime of our tool is similar with that of $if - g$; $iresyn$ ABC script.

All MIG output Verilog files passed formal verification tests (ABC *cec* and Synopsys Formality) with success.

3.6.4 ASIC Results

Table 3.3 shows the results for ASIC design (synthesis followed by place and route) at a commercial 22-nm technology node.[2] In total, we see that by using *MIGhty* as front-end to the ASIC design flow, we obtained better final circuits, in all relevant metrics including area, delay and power. For the delay, which was our critical design constraint, we observe an improvement of about 13 %. This improvement is not as large as the one we saw at the logic optimization level because some of the gain got absorbed by the interconnect overhead during physical design. However, we still see a coherent trend: We obtained 4 and 3 % reductions in area and power.

3.6.5 FPGA Results

Table 3.4 shows the results for FPGA design (synthesis followed by place and route) on a commercial 28-nm technology node[3]. By employing *MIGhty* as front-end to the FPGA design flow, we obtain better final circuits, in LUT count, delay and power metrics. For the delay, that was our critical design constraint, we observe an improvement of about 10 %. Also here, P&R absorbs part of the advantage predicted at the logic-level. Regarding LUT number and power, we see improvements of about 10 and 5 %, respectively. Some of the values reported (marked by*) are just post synthesis results because the placement and routing on FPGA failed due to excessive number of I/Os.

---

[2]Design tools and library names cannot be disclosed due to our license agreement.

Table 3.3 MIG 22-nm ASIC design results

Benchmark	MIGhty+ASIC flow			ASIC flow		
	μm^2	ns	mW	μm^2	ns	mW
DSP	6958.23	**0.57**	1.82	1841.76	**2.95**	1.82
ac97_ctrl	2045.48	**0.12**	0.55	2070.83	**0.15**	0.56
aes_core	4599.62	**0.29**	1.75	4417.46	**0.29**	1.64
des_area	853.21	**0.31**	0.59	1084.60	**0.36**	0.53
des_perf	14417.90	**0.20**	11.21	15808.09	**0.23**	11.81
ethernet	10835.31	**0.25**	1.61	10631.93	**0.29**	1.59
i2c	210.13	**0.10**	0.04	210.04	**0.11**	0.04
mem_ctrl	1359.41	**0.30**	0.27	1372.58	**0.33**	0.27
pci_b32	3215.69	**0.26**	0.79	3259.76	**0.29**	0.79
pci_spoci	159.34	**0.16**	0.08	177.47	**0.16**	0.09
sasc	125.12	**0.08**	0.02	139.98	**0.10**	0.02
simple_spi	169.60	**0.12**	0.04	178.64	**0.14**	0.04
spi	542.22	**0.39**	0.21	503.41	**0.42**	0.18
ss_pcm	85.33	**0.08**	0.02	89.23	**0.08**	0.02
systemcaes	1328.08	**0.35**	0.65	1427.94	**0.43**	0.66
systemcdes	538.97	**0.31**	0.37	641.30	**0.33**	0.45
tv80	1299.34	**0.43**	0.37	1213.84	**0.50**	0.40
usb_funct	2269.22	**0.25**	0.72	2337.65	**0.26**	0.77
usb_phy	111.15	**0.05**	0.02	115.73	**0.07**	0.02
MUL32	1862.55	**0.55**	1.81	1748.45	**0.56**	1.90
sqrt32	498.65	**2.54**	0.62	504.76	**2.74**	0.62
diffeq1	3460.48	**3.19**	4.33	3713.87	**3.49**	4.68
div16	595.86	**1.64**	0.26	948.66	**2.06**	0.40
hamming	325.65	**0.90**	0.56	348.46	**1.04**	0.58
MAC32	2281.57	**0.58**	1.95	2194.88	**0.60**	1.89
metric_c	4274.04	**1.36**	1.68	4642.09	**1.55**	1.72
revx	1401.04	**2.23**	1.42	1451.11	**2.63**	1.48
mul64	6378.20	**1.43**	7.01	6330.08	**1.82**	6.95
max	628.23	**0.45**	0.33	631.46	**0.56**	0.33
square	4031.05	**0.46**	3.69	3895.13	**0.67**	3.57
log2	6784.70	**3.07**	7.45	7197.50	**3.59**	8.03
Total	83645.37	**23.02**	53.37	86270.09	**26.47**	55.04

In summary, MIG synthesis technology enables a consistent advantage over the state-of-the-art commercial design flows. It is worth noticing that we employed MIG optimization just as a front-end to an existing commercial flow. We foresee even better results by integrating MIG optimization inside the synthesis engine of commercial tools.

Table 3.4 MIG 28-nm FPGA Design Results

Benchmark	MIGhty+FPGA flow			FPGA flow		
	LUT6	ns	W	LUT6	ns	W
DSP*	9599	**8.22**	7.76	9501	**8.54**	7.73
ac97_ctrl*	2417	**4.54**	3.91	2444	**4.67**	3.92
aes_core	4440	**5.54**	1.93	4788	**5.63**	1.94
des_area	955	**15.24**	0.96	1212	**15.73**	0.98
des_perf*	8480	**5.22**	18.56	11350	**5.40**	18.75
etherne*t	14840	**6.26**	23.89	16343	**6.74**	23.84
i2c	274	**10.58**	0.83	264	**10.38**	0.83
mem_ctrl*	1929	**6.74**	2.00	2044	**7.25**	1.99
pci_b32*	4542	**5.76**	7.77	4741	**6.39**	7.78
pci_spoci	260	**9.86**	0.81	290	**9.99**	0.81
sasc	141	**10.02**	0.88	137	**10.04**	0.88
simple_spi	192	**9.91**	0.93	200	**10.23**	0.93
spi	994	**15.72**	1.32	814	**18.57**	1.35
ss_pcm	92	**9.60**	0.78	89	**9.58**	0.78
systemcaes	1445	**6.67**	2.31	1445	**6.96**	2.32
systemcdes	667	**14.93**	1.31	798	**15.90**	1.31
tv80	1892	**16.44**	1.57	1975	**17.47**	1.57
usb_funct*	2988	**6.02**	3.25	2887	**5.79**	3.21
usb_phy	97	**10.00**	0.82	94	**10.06**	0.82
MUL32	1776	**11.05**	0.88	1867	**12.02**	0.89
sqrt32	447	**25.46**	0.68	560	**27.81**	0.70
diffeq1	5134	**22.36**	1.56	6545	**30.89**	1.82
div16	1160	**26.03**	0.72	765	**28.12**	0.70
hamming	519	**16.20**	13.16	657	**17.65**	17.81
MAC32	2220	**12.47**	0.93	2338	**15.83**	0.94
metric_c	5486	**34.57**	1.11	6416	**38.65**	1.13
revx	2010	**26.19**	0.79	2333	**31.04**	0.80
mul64	7109	**22.54**	1.77	6224	**25.07**	1.41
max	952	**20.10**	1.04	754	**22.19**	1.04
square	4327	**17.05**	1.16	3579	**17.56**	1.11
log2	9944	**44.13**	1.42	14166	**51.75**	1.79
Total	97328	**455.41**	106.81	107620	**503.97**	111.88

3.7 MIGs as Native Design Abstraction
for Nanotechnologies

MIGs are the natural and native design abstraction for several emerging technologies
where the circuit primitive is a majority voter, rather than a switch. In this section,
we test the efficacy of MIGs in the synthesis of spin-wave devices and resistive RAM
nanotechnologies. We start by introducing general notions on these two nanotechnologies in order to explain their primitive logic operation. Then, we show how the
MIG logic model fits and actually helps in exploiting at best the expressive power of
the considered nanotechnologies.

Note that many other nanodevices may benefit from the presented majority synthesis methodologies [46, 47].

3.7.1 MIG-Based Synthesis

MIGs enable compact logic representation and powerful logic optimization. They
already show very promising results for traditional CMOS technology [3, 4]. Moreover, if the target technology natively realizes the MIG primitive function, i.e., a
majority voter, the use of MIGs in circuit synthesis produces superior results. We
use *MIGhty* to synthesize circuits in voting-intrinsic nanotechnologies.

Depending on the target nanotechnology, we either use MIGs for a direct one-to-
one mapping into nanodevices or as a frontend to a standard synthesis tool. In both
cases, no pre-partitioning is strictly required as MIG are not canonical per se, thus
they scale efficiently with the design size.

More details on MIG-based synthesis are given for each specific nanotechnology.

3.7.2 Spin-Wave Devices

Spin Wave Devices (SWDs) are digital devices where information transmission happens via spin waves instead of conventional carriers (electrons and holes). The SWD
physical mechanism enables ultra-low power operation, almost two orders of magnitude lower than the one of state of the art CMOS [49].

SWDs operate via propagated oscillation of the magnetization in an ordered
magnetic material [50]. That oscillation (spin wave) is generated, manipulated and
detected though a synthetic multi-ferroic component, commonly called *Magneto-
Electric* (ME) cell [51]. The characteristic size of spin-wave devices is the spin
wavelength, whose values may range from 30 up to 200 nm [49].

On top of the extremely low power consumption of SWD logic, which is a
key technological asset, the employment of wave computation in digital circuits
can enhance its logic expressive power. SWD logic computation is based on the

Fig. 3.6 Primitive gate areas and designs for SWD technology. All distances are parameterized with the spin wave wavelength λ_{SW} [48]

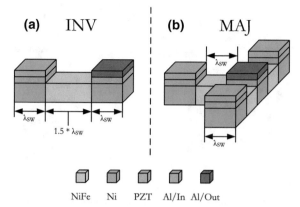

Table 3.5 Cost functions for MIGs mapped onto SWDs

MIG element	SWD gate	Area cost	Delay cost
Majority node	Majority gate	4	1
Complemented edge	Inverter gate	1	1

interference of spin waves. Depending on the phase of the propagating spin waves/signals, their interference is constructive or destructive. The final interference result is translated to the output via magneto-electric cells. In this scenario, an inverter is simply a waveguide with length equal to $1.5\times$ of the spin wavelength (λ_{SW}). In this way, the information encoded in the phase of the SW signal arrives inverted to the output ME cell, Fig. 3.6a. The actual logic primitive in SWD technology is the majority voter, which is implemented by the symmetric merging of three waveguides Fig. 3.6b. Here, the length of each waveguide is $1.0\times$ the spin wavelength. In the majority voter structure, the spin wave signal at the output is determined by the majority phase of the input spin waves.

In order to fully exploit the SWD technology potential, we have to leverage the native logic primitives spin wave logic offer. In SWDs, the logic primitive is a majority voter. Standard synthesis techniques are inadequate to harness this potential. However, the novel MIG data structure previously introduced naturally matches the voting functionality of SWD logic. For this reason, we use MIGs to represent and synthesize SWD circuits. The intrinsic correspondence between MIG elements and SWDs makes MIG optimization naturally extendable to obtain minimum cost SWD implementations. For this purpose, ad hoc cost functions are assigned to MIG elements during optimization as per Table 3.5. These cost functions are derived from the SWD technology implementation of majority and inverter gates in Fig. 3.6.

For the sake of clarity, we comment on our proposed MIG-based SWD synthesis flow by means of an example. The objective function in this example is $g = x \cdot (y + u \cdot v)$. This function is initially represented by the MIG in Fig. 3.7(left), which has a SWD delay cost of 4 and an SWD area cost of 14. By using Ω transformations, it is

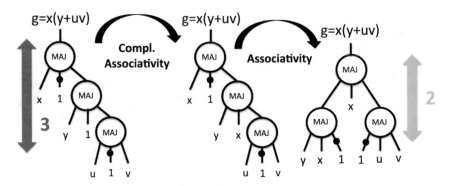

Fig. 3.7 Optimization of the MIG representing the function $g = x \cdot (y + u \cdot v)$. Initial MIG counts 3 nodes and 3 levels. Final MIG counts 3 nodes and 2 levels

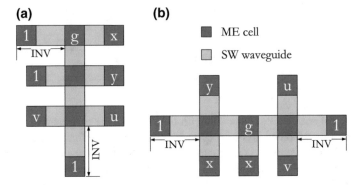

Fig. 3.8 SWD circuit implementing function g, **a** from example in Fig. 3.3 (*left*). **b** From example in Fig. 3.3 (*right*) which is optimized in size and depth

possible to reach the optimized MIG depicted by Fig. 3.7(right). Such an optimized MIG counts the same number of nodes and complemented edges of the original one but one fewer level of depth. In this way, the associated area cost remains 14 but the delay is reduced to 3. After the optimization, each MIG element is mapped onto its corresponding SWD gate. Figure 3.8 depicts the SWD mapping for the original (a) and optimized (b) MIGs.

As one can visually notice, the circuit in Fig. 3.8b features roughly the same area occupation as the one in Fig. 3.8a but shorter input-output path. Following the theoretical cost functions employed, the achieved speed-up is roughly 25 %. Including the physical models and assumptions presented in [48], the refined speed-up becomes 18.2 %.

We validate hereafter the efficiency of our proposed MIG-based SWD synthesis flow for larger circuits [52]. We also provide a comparison reference to 10-nm CMOS technology.

In MIG-based SWD synthesis, we employed the *MIGhty* MIG optimizer [3]. As traditional-synthesis counterpart, we employed ABC tool [27] with optimization commands *resyn2* and produced in output an *AND-Inverter Graph* (AIG). The AIGs mapping procedure onto SWDs is in common with MIGs: AND nodes are simply mapped to MAJ gates with one input biased to logic 0. For advanced CMOS, we used a commercial synthesis tool fed with a standard-cell library produced by a 10-nm CMOS process flow. The circuit benchmarks are taken from the MCNC suite.

The cost functions for MIG optimization are taken from Table 3.5. In addition to the direct cost of SWD gates, our design setup takes also into consideration the integration in a VLSI environment given input and output overhead, as presented in [52]. The final synthesis values presented hereafter are comprising all these costs.

Table 3.6 shows all synthesis results for SWD and CMOS technologies. We summarize in Table 3.7 the performance of the benchmarks in the *Area-Delay-Power* (ADP) product to better point out the significant improvement MIG synthesis brings to light. SWD circuits synthesized via MIGs have 1.30× smaller ADP than SWD circuits synthesized via traditional AIGs. This is thanks to the SWD delay improvement enabled by MIGs. As compared to the 10-nm CMOS technology, SWD circuits synthesized by MIGs have 17.02× smaller ADP, offering an ultra-low power, compact SWD implementation with reduced penalty in delay.

Results showed that MIG synthesis naturally fits SWD technology needs. Indeed, MIGs enhanced SWD strengths (area and power) and alleviated SWD issues (delay).

3.7.3 Resistive RAM

Multitude of emerging *Non-Volatile Memories* (NVM) are receiving widespread research attention as candidates for high-density and low-cost storage. NVMs store information as an internal resistive state, which can be either a *Low Resistance State* (LRS) or a *High Resistance State* (HRS) [53]. Among the different types of NVMs, Redox-based *Resistive RAM* (RRAM) is considered a leading candidate due to its high density, good scalability, low power and high performance [54, 55]. A different and arguably more tantalizing aspect of RRAMs is their ability to do primitive Boolean logic. The possibility of in-memory computing significantly widens the scope of the commercial applications. To undertake a logic computation, RRAM-based switches are needed. *Bipolar Resistive Switches* (BRS) [56] and *Complementary Resistive Switches* (CRS) [57] have been presented for this purpose. BRS and CRS are devices with two terminals, denoted P and Q. BRS can be used in ultra-dense passive crossbar arrays but suffer from the formation of parasitic currents which create sneak paths. This problem can be alleviated by constructing a CRS device, which connects two BRS anti-serially [57]. For the sake of clarity, we report in Fig. 3.9 the CRS device conceptual structure proposed in [57] and its

Table 3.6 Experimental results for SWDs-MIG Synthesis

Benchmarks	I/O	SWD technology—MIG			SWD technology—AIG			CMOS technology—commercial tool		
		A (μm^2)	D (ns)	P (μW)	A (μm^2)	D (ns)	P (μW)	A (μm^2)	D (ns)	P (μW)
C1355	41/32	16.95	5.81	0.12	13.88	5.81	0.10	36.27	0.39	68.06
C1908	33/25	16.13	7.30	0.09	12.81	7.9	0.07	32.68	0.53	61.19
C6288	32/32	77.57	26.05	0.12	70.93	28.43	0.11	131.94	1.32	425.21
bigkey	487/421	152.50	3.14	2.11	170.99	3.14	2.34	238.85	0.32	262.50
my_adder	33/17	9.42	6.11	0.07	5.00	10.28	0.04	17.83	0.44	23.94
cla	129/65	36.57	7.60	0.21	32.21	11.77	0.19	72.49	0.62	88.48
dalu	75/16	50.47	6.71	0.31	39.17	9.39	0.25	46.59	0.36	34.63
b9	41/21	5.60	2.24	0.08	5.60	2.54	0.08	5.92	0.09	4.73
count	35/16	6.36	2.54	0.11	4.67	6.11	0.09	8.90	0.32	6.56
alu4	14/8	47.81	4.62	0.42	49.22	4.62	0.43	87.20	0.34	72.39
clma	416/115	433.59	12.96	1.37	450.15	14.15	1.42	231.69	0.51	177.82
mm30a	124/120	41.57	30.52	0.06	35.70	37.66	0.05	68.40	1.68	47.19
s38417	1494/1571	319.86	7.01	1.92	319.86	7.9	1.88	609.94	0.53	740.73
misex3	14/14	45.84	4.33	0.43	44.14	4.62	0.41	78.02	0.26	59.34
Average	212/176	90.02	9.07	0.53	89.60	11.02	0.53	119.05	0.55	148.06

Table 3.7 Summarizing performance results of SWD and CMOS technologies

Technology	ADP product (a.u.)	Gain versus CMOS	Gain versus AIG
CMOS	9707.06	–	–
SWD—AIG	526.25	18.45 ×	–
SWD—MIG	432.59	22.44 ×	1.22×

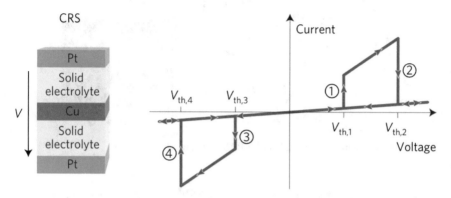

Fig. 3.9 CRS conceptual structure and sweep properties from [57]

sweep properties. Their internal resistance state of the device, Z, can be modified by applying a positive or a negative voltage V_{PQ}. The functionality of BRS/CRS can be summarized by a state machine, as shown in Fig. 3.10. Further details can be found in [58]. Transition occurs only for the conditions $P = 0$, $Q = 1$, i.e., $V_{PQ} < 0$ so $Z \rightarrow 0$ and $P = 1$, $Q = 0$, i.e., $V_{PQ} > 0$ so $Z \rightarrow 1$. By denoting Z as the value stored in the switch and Z_n the value stored after the application of signals on P and Q, it is possible to express Z_n as the following:

$$Z_n = (P.\overline{Q}).\overline{Z} + (P + \overline{Q}).Z$$
$$= P.Z + \overline{Q}.Z + P.\overline{Q}.\overline{Z}$$
$$= P.Z + \overline{Q}.Z + P.\overline{Q}.Z + P.\overline{Q}.\overline{Z}$$
$$= P.Z + \overline{Q}.Z + P.\overline{Q}$$
$$= M_3(P, \overline{Q}, Z)$$

where M_3 is the majority Boolean function with 3 inputs.

The aforementioned resistive RAM technology enables a in-memory computing system, which exploits memristive devices to perform both standard storage and computing operations, such as majority voting.

The possibility of in-memory computing for RRAM technology can increase the intelligence of many portable electronic devices. However, to fully exploit this opportunity, the primitive Boolean operation in RRAM technology needs to be fully understood and natively handled by design tools. In this context, the MIG data structure offers a native logic abstraction for RRAM in-memory computation. To demonstrate the efficacy of the RRAM-MIG coupling, we map a lightweight cryptography block cipher [60] on a RRAM array using MIG-based design techniques [59].

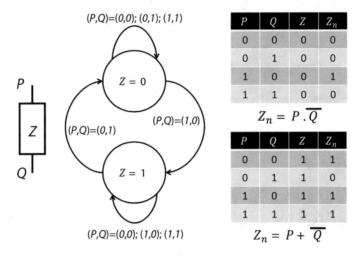

P	Q	Z	Z_n
0	0	0	0
0	1	0	0
1	0	0	1
1	1	0	0

$$Z_n = P \cdot \overline{Q}$$

P	Q	Z	Z_n
0	0	1	1
0	1	1	0
1	0	1	1
1	1	1	1

$$Z_n = P + \overline{Q}$$

Fig. 3.10 Resistive majority operation with BRS/CRS devices [59]

The target cryptography block cipher is PRESENT, originally introduced in [60]. We briefly review its encryption mechanism hereafter.

3.7.3.1   PRESENT Encryption

A PRESENT encryption consists of 31 rounds, through which multiple operations are performed on the 64-bit plaintext and finally produces a 64-bit ciphertext. The rounds modify the plaintext, which is referred as STATE internally. The operation of the cipher components are *addRoundKey*, *sBoxLayer*, *pLayer*, and *KeyUpdate* [60].

For the sake of brevity, we give here details only on the *sBoxLayer* operation. We refer the interested reader to [60] for details on the other operations. The *sBoxLayer* operation divides the 64-bit word into 16 parts of 4-bit each. Each 4-bit portion is the processed individually by a 4-input, 4-output combinational Boolean function, called operator S. In order to map S into the RRAM memory array, we use MIG representation and optimization. The optimization goal is to reduce the number of majority operations.

3.7.3.2   S Operator Mapping

The S operator is nothing but a Boolean function with primary inputs pi_0, pi_1, pi_2, pi_3 and primary outputs po_0, po_1, po_2, po_3. For the sake of brevity, we only represent in Fig. 3.11 the MIG representation for po_0 that consists of 11 majority nodes. Then, each majority node is mapped into a set of primitive RRAM memory/computing instructions. For instance, the portion highlighted in grey on the net-

Fig. 3.11 MIG representing the output po_0 in the S encryption operator

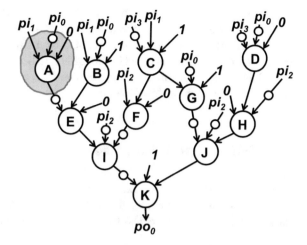

Table 3.8 Experimental results for RRAM-MIG synthesis PRESENT implementation performances

Operation	Instructions ($\#M_3$)	Cycles ($\#R/W$)
Key copy	80	720
Cipher copy	64	576
AddRoundKey	448	4032
sBoxLayer	608	5472
pLayer	64	576
KeyUpdate	760	6840
	Instructions	Cycles
PRESENT block	58 872	455 184
	Energy (pJ)	Throughput (kbps)
PRESENT block	5.88	120.7

work corresponds to the operation $M(pi_1, pi_0, 0)$. Assuming that logic 0 is the previous value stored in the array, an immediate majority instruction computes the corresponding portion of logic. The total S operator requires a total of 38 cycles for its operation in the RRAM array.

Using an analogous MIG-mapping approach, all the PRESENT encryption operations can be performed directly on the RRAM array.

The overall performance of the MIG-based PRESENT implementation on the RRAM array has been estimated considering a RRAM technology aligned with the ITRS 2013 predictions. More precisely, we assume a write time of 1 ns and a write energy of 0.1 fJ/bit. Table 3.8 summarizes the number of M_3 instructions and *Read/Write* (R/W) cycles required by the different operations of the PRESENT cipher.

The total number of primitive majority instructions for the encryption of a 64-bit cipher text is 58872 [59]. The total throughput reachable by the system is 120.7

kbps, making it comparable to silicon implementations [60]. Finally, the total energy required for one block encryption operation is 5.88 pJ.

This remarkable design result is enabled by a strong MIG optimization on the critical logic operations involved in PRESENT. Otherwise, its implementation without MIGs would require many more primitive RM_3 instructions making it inefficient when compared to the state-of-the-art.

3.8 Extension to MAJ-*n* Logic

In this section, we extend the axiomatization of MAJ-3 logic to MAJ-*n* logic. First, we show the axiomatization validity in the Boolean domain. Then, we demonstrate the axiomatization completeness by inclusion of other complete Boolean axiomatizations.

3.8.1 Generic MAJ-n/INV Axioms

The five axioms for MAJ-3/INV logic presented in Sect. 3.3.2 deal with *commutativity*, *majority*, *associativity*, *distributivity*, and *inverter propagation* laws. The set of equations in Eq. 3.4 extends their domain to an arbitrary odd number n of variables.

$$\Omega_n \begin{cases} \textbf{Commutativity} : \Omega_n.C \\ M_n(x_1^{i-1}, x_i, x_{i+1}^{j-1}, x_j, x_{j+1}^n) = M_n(x_1^{i-1}, x_j, x_{i+1}^{j-1}, x_i, x_{j+1}^n) \\ \textbf{Majority} : \Omega_n.M \\ \text{If}(\lceil \frac{n}{2} \rceil \text{ elements of } x_1^n \text{ are equal to } y) : \\ \quad M_n(x_1^n) = y \\ \text{If } (x_i \neq x_j) : M_n(x_1^n) = M_{n-2}(y_1^{n-2}) \\ \quad \text{where } y_1^{n-2} = x_1^n \text{ removing}\{x_i, x_j\} \\ \textbf{Associativity} : \Omega_n.A \\ M_n(z_1^{n-2}, y, M_n(z_1^{n-2}, x, w)) = M_n(z_1^{n-2}, x, M_n(z_1^{n-2}, y, w)) \\ \textbf{Distributivity} : \Omega_n.D \\ M_n(x_1^{n-1}, M_n(y_1^n)) = \\ \\ M_n(M_n(x_1^{n-1}, y_1), M_n(x_1^{n-1}, y_2), \ldots, M_n(x_1^{n-1}, y_{\lceil \frac{n}{2} \rceil}), y_{\lceil \frac{n}{2} \rceil+1}, \ldots, y_n) = \\ \\ M_n(M_n(x_1^{n-1}, y_1), M_n(x_1^{n-1}, y_2), \ldots, M_n(x_1^{n-1}, y_{\lceil \frac{n}{2} \rceil+1}), y_{\lceil \frac{n}{2} \rceil+2}, \ldots, y_n) = \\ \\ M_n(M_n(x_1^{n-1}, y_1), M_n(x_1^{n-1}, y_2), \ldots, M_n(x_1^{n-1}, y_{n-1}), y_n) \\ \textbf{Inverter Propagation} : \Omega_n.I \\ M_n(x_1^n)' = M_n(x_1^{n\prime}) \end{cases}$$

$$(3.4)$$

Commutativity means that changing the order of the variables in M_n does not change the result. Majority defines a logic decision threshold and a hierarchical reduction of majority operators with complementary variables. Associativity says that swapping pairs of variables between cascaded M_n sharing $n - 2$ variables does not change the result. In this context, it is important to recall that $n - 2$ is an odd number if n is an odd number. Distributivity delimits the re-arrangement freedom of variables over cascaded M_n operators. Inverter propagation moves complementation freely from the outputs to the inputs of a M_n operator, and *viceversa*.

For the sake of clarity, we give an example for each axiom over a finite n-arity. Commutativity with $n = 5$:
$$M_5(a, b, c, d, e) = M_5(b, a, c, d, e) = M_5(a, b, c, e, d).$$
Majority with $n = 7$:
$$M_7(a, b, c, d, e, g, g') = M_5(a, b, c, d, e).$$
Associativity with $n = 5$:
$$M_5(a, b, c, d, M_5(a, b, c, g, h)) = M_5(a, b, c, g, M_5(a, b, c, d, h)).$$
Distributivity with $n = 7$:
$$M_7(a, b, c, d, e, g, M_7(x, y, z, w, k, t, v)) = M_7(M_7(a, b, c, d, e, g, x), M_7(a, b, c, d, e, g, y), M_7(a, b, c, d, e, g, z), M_7(a, b, c, d, e, g, w), k, t, v).$$
Inverter propagation with $n = 9$:
$$M_9(a, b, c, d, e, g, h, x, y)' = M_9(a', b', c', d', e', g', h', x', y').$$

3.8.2 Soundness

To demonstrate the validity of these laws, and thus the validity of the MAJ-n axiomatization, we need to show that each equation in Ω_n is sound with respect to the original domain, i.e., $(\mathbb{B}, M_n,', 0, 1)$. The following theorem addresses this requirement.

Theorem 3.9 *Each axiom in Ω_n is sound (valid) w.r.t. $(\mathbb{B}, M_n,', 0, 1)$.*

Proof We prove the soundness of each axiom separately.

Commutativity $\Omega_n.C$ Since majority is defined on reaching a threshold $\lceil n/2 \rceil$ of true inputs then it is independent of the order of its inputs. This means that changing the order of operands in M_n does not change the output value. Thus, this axioms is valid in $(\mathbb{B}, M_n,', 0, 1)$.

Majority $\Omega_n.M$ Majority first defines the output behavior of M_n in the Boolean domain. Being a definition, it does not need particular proof for soundness. Consider then the second part of the majority axiom. The hierarchical inclusion of M_{n-2} derives from the mutual cancellation of complementary variables. In a binary majority voting system of n electors, two electors voting to opposite values annihilate theirselves. The final decision is then just depending on the votes from the remaining $n - 2$ electors. Therefore, this axiom is valid in $(\mathbb{B}, M_n,', 0, 1)$.

Associativity $\Omega_n.A$ We split this proof in three parts that cover the whole Boolean space. Thus, it is sufficient to prove the validity of the associativity axiom for each of this parts. **(1) the vector z_1^{n-2} contains at least one logic 1 and one logic 0.** In this case, it is possible to apply $\Omega_n.M$ and reduce M_n to M_{n-2}. If we remain in case (1), we can keep applying $\Omega_n.M$. At some point, we will end up in case (2) or (3). **(2) the vector z_1^{n-2} contains all logic 1.** For $n > 3$, the final voting decision is 1 for both equations, so the equality holds. In case $n = 3$, the majority operator collapses into a disjunction operator. Here, the validity of the associativity axiom follows then from traditional disjunction associativity. **(3) the vector z_1^{n-2} contains all logic 0.** For $n > 3$, the final voting decision is 0 for both equations, so the equality holds. In case $n = 3$, the majority operator collapses into a conjunction operator. Here, the validity of the associativity axiom follows then from traditional conjunction associativity.

Distributivity $\Omega_n.D$ We split this proof in three parts that cover the whole Boolean space. Thus, it is sufficient to prove the validity of the distributivity axiom for each of this parts. Note that the distributivity axiom deals with a majority operator M_n where one inner variable is actually another independent majority operator M_n. Distributivity rearranges the computation in M_n moving up the variables at the bottom level and down the variables at the top level. In this part of the proof we show that such rearrangement does not change the functionality of M_n, i.e., the final voting decision in $\Omega_n.D$. Recall that n is an odd integer greater than 1 so $n - 1$ must be an even integer. **(1) half of x_1^{n-1} values are logic 0 and the remaining half are logic 1.** In this case, the final voting decision in axiom $\Omega_n.D$ only depends on y_1^n. Indeed, all elements in x_1^{n-1} annihilate due to axiom $\Omega_n.M$. In the two identities of $\Omega_n.D$, we see that when x_1^{n-1} annihilate the equations simplify to $M_n(y_1^n)$, according to the predicted behavior. **(2) at least $\lceil n/2 \rceil$ of x_1^{n-1} values are logic 0.** Owing to $\Omega_n.M$, the final voting decision in this case is logic 0. This is because more than half of the variables are logic 0 matching the prefixed voting threshold. In the two identities of $\Omega_n.D$, we see that more than half of the inner M_n evaluate to logic 0 by direct application of $\Omega_n.M$. In the subsequent phase, also the outer M_n evaluates to logic 0, as more than half of the variables are logic 0, according to the predicted behavior. **(3) at least $\lceil n/2 \rceil$ of x_1^{n-1} values are logic 1.** This case is symmetric to the previous one.

Inverter Propagation $\Omega_n.I$ Inverter propagation moves complementation from output to inputs, and *viceversa*. This axiom is a special case of the self-duality property previously presented. It holds for all majority operators in $(\mathbb{B}, M_n, ', 0, 1)$. ∎

The soundness of Ω_n in $(\mathbb{B}, M_n, ', 0, 1)$ guarantees that repeatedly applying Ω_n axioms to a Boolean formula we do not corrupt its original functionality. This property is of interest in logic manipulation systems where functional correctness is an absolute requirement.

3.8.3 Completeness

While soundness speaks of the correctness of a logic systems, completeness speaks of its manipulation capabilities. For an axiomatization to be complete, all possible manipulations of a Boolean formula must be attainable by a sequence, possibly long, of primitive axioms.

We study the completeness of Ω_n axiomatization by comparison to other complete axiomatizations of Boolean logic. The following theorem shows our main result.

Theorem 3.10 *The set of five axioms in Ω_n is complete w.r.t. $(\mathbb{B}, M_n,' , 0, 1)$.*

Proof We first recall that Ω_3 is complete w.r.t. $(\mathbb{B}, M_3,' , 0, 1)$ as proved by Theorem 3.6. We consider then Ω_n. First note that $(\mathbb{B}, M_n,' , 0, 1)$ naturally includes $(\mathbb{B}, M_3,' , 0, 1)$. Similarly, Ω_n axioms inherently extend the ones in Ω_3. Thus, the completeness property is inherited provided that Ω_n axioms are sound. However, Ω_n soundness is already proven in Theorem 3.9. Thus, Ω_n axiomatization is also complete. ∎

Being sound and complete, the axiomatization Ω_n defines a consistent framework to operate on Boolean logic via majority operators. It also gives directions for future applications of majority/inverters in computer science, such as Boolean satisfiability, repetition codes, threshold logic, artificial neural network etc.

3.9 Summary

In this chapter, we proposed a paradigm shift in representing and optimizing logic circuits, by using only majority (MAJ) and inversion (INV) as basic operations. We presented the *Majority-Inverter Graphs* (MIGs): a directed acyclic graph consisting of three-input majority nodes and regular/complemented edges. We developed algebraic and Boolean optimization techniques for MIGs and we embedded them into a tool, called *MIGhty*. Over the set of IWLS'05 (arithmetic intensive) benchmarks, *MIGhty* enabled a 7% (16%) depth reduction in LUT-6 circuits mapped by ABC while also reducing size and power activity, with respect to similar AIG optimization. Employed as front-end to a delay-critical 22-nm ASIC flow, *MIGhty* reduced the average delay/area/power by about 13%/4%/3%, over 31 benchmarks. We also demonstrated improvements in delay/area/power by 10%/10%/5% for a commercial 28-nm FPGA flow. Results on two emerging nanotechnologies, i.e., spin-wave devices and resistive RAM, demonstrated that MIGs are essential to permit a fair technology evaluation where the logic primitive is a majority voter. Finally, we extended the axiomatization of MAJ-3 logic to MAJ-*n* logic, with *n* odd, preserving soundness and completeness properties.

References

1. G. De Micheli, *Synthesis and Optimization of Digital Circuits* (McGraw-Hill, New York, 1994)
2. N. Song et al., EXORCISM-MV-2: minimization of ESOP expressions for MV input incompletely specified functions, in *Proceedings on MVL* (1993)
3. L. Amarú et al., Majority-inverter graph: a novel data-structure and algorithms for efficient logic optimization, in *Proceedings of the DAC'14*
4. L. Amarú et al., Boolean logic optimization in majority-inverter graph, in *Proceedings of the DAC'15*
5. T. Sasao, *Switching Theory for Logic Synthesis* (Springer, New York, 1999)
6. R.L. Rudell, A. Sangiovanni-Vincentelli, Multiple-valued minimization for PLA optimization. IEEE TCAD **6**(5), 727–750 (1987)
7. R.E. Bryant, Graph-based algorithms for boolean function manipulation. IEEE TCOMP **C-35**(8), 677–691 (1986)
8. R. Brayton, A. Mishchenko, ABC: an academic industrial-strength verification tool, in *Proceedings of the CAV* (2010)
9. E.J. McCluskey, Minimization of boolean functions. Bell Syst. Tech. J. **35**(6), 1417–1444 (1956)
10. R.K. Brayton, Multilevel logic synthesis. Proc. IEEE **78**(2), 264–300 (1990)
11. R.K. Brayton et al., MIS: a multiple-level logic optimization system. IEEE Trans. CAD **6**(6), 1062–1081 (1987)
12. R.K. Brayton et al., The decomposition and factorization of boolean expressions, in *Proceedings of the ISCAS'82*
13. R.K. Brayton et al., MIS: a multiple-level logic optimization system. IEEE TCAD **6**(6), 1062–1081 (1987)
14. A. Mishchenko et al., Using simulation and satisfiability to compute flexibilities in boolean networks. IEEE TCAD **25**(5), 743–755 (2006)
15. S.C. Chang et al., Perturb and simplify: multilevel boolean network optimizer. IEEE TCAD **15**(12), 1494–1504 (1996)
16. S.C. Chang, L.P. Van Ginneken, M. Marek-Sadowska, Circuit optimization by rewiring. IEEE TCOMP **48**(9), 962–970 (1999)
17. R. Ashenhurst, The decomposition of switching functions, in *Proceedings of the International Symposium on the Theory of Switching*, pp. 74–116 (1957)
18. J.P. Roth, R.M. Karp, Minimization over boolean graphs. IBM J. 661–664 (1962)
19. H.A. Curtis, *A New Approach to the Design of Switching Circuits* (Van Nostrand, Princeton, 1962)
20. C. Yang, M. Ciesielski, BDS: a BDD-based logic optimization system. IEEE TCAD **21**(7), 866–876 (2002)
21. N. Vemuri et al., BDD-based logic synthesis for LUT-based FPGAs. ACM TODAES **7**(4), 501–525 (2002)
22. L. Amaru, P.-E. Gaillardon, G. De Micheli, BDS-MAJ: a BDD-based logic synthesis tool exploiting majority decomposition, in *Proceedings of the DAC* (2013)
23. V. Bertacco, M. Damiani, Disjunctive decomposition of logic functions, in *Proceedings of the ICCAD'97*, pp. 78–82
24. A. Mishchenko, R. Brayton, Faster logic manipulation for large designs, in *Proceedings of the IWLS'13*
25. A. Mishchenko, S. Chatterjee, R. Brayton, DAG-aware AIG rewriting a fresh look at combinational logic synthesis, in *Proceedings of the DAC* (2006)
26. A. Mishchenko, R. Brayton, Scalable logic synthesis using a simple circuit structure, in *Proceedings of the IWLS* (2006)
27. ABC synthesis tool. http://www.eecs.berkeley.edu/~alanmi/abc/
28. H.S. Miller, R.O. Winder, Majority-logic synthesis by geometric methods. IRE Trans. Electron. Comput. **EC–11**, 89–90 (1962)

29. Y. Tohma, Decompositions of logical functions using majority decision elements. IEEE Trans. Electron. Comput. **EC–13**, 698–705 (1964)
30. F. Miyata, Realization of arbitrary logical functions using majority elements. IEEE Trans. Electron. Comput. **EC–12**, 183–191 (1963)
31. E.V. Huntington, Sets of independent postulates for the algebra of logic. Trans. Am. Math. Soc. **5**(3), 288–309 (1904)
32. B. Jonsson, Boolean algebras with operators. Part I. Am. J. Math. **73**, 891–939 (1951)
33. G. Birkhoff, *Lattice Theory* (American Mathematical Society, New York, 1967)
34. J.R. Isbell, Median algebra. Trans. Am. Math. Soc. **260**, 319–362 (1980)
35. G. Birkhoff, A ternary operation in distributive lattices. Bull. Am. Math. Soc. **53**(1), 749–752 (1947)
36. D. Knuth, *The Art of Computer Programming*, vol. 4A, Part 1 (Addison-Wesley, New Jersey, 2011)
37. M. Krause et al., On the computational power of depth-2 circuits with threshold and modulo gates. Theor. Comput. Sci. **174**(1), 137–156 (1997)
38. S. Muroga et al., The transduction method-design of logic networks based on permissible functions. IEEE TCOMP **38**(10), 1404–1424 (1989)
39. R.E. Lyons, W. Vanderkulk, The use of triple-modular redundancy to improve computer reliability. IBM J. Res. Dev. **6**(2), 200–209 (1962)
40. A.C. Gomes et al., Methodology for achieving best trade-off of area and fault masking coverage in ATMR. IEEE LATW (2014)
41. M. Pedram, Power minimization in IC design: principles and applications. ACM TODAES **1**(1), 3–56 (1996)
42. M. Parnas et al., Proclaiming dictators and juntas or testing boolean formulae, *Combinatorial Optimization* (Springer, New York, 2001), pp. 273–285
43. http://lsi.epfl.ch/MIG
44. A. Mishchenko et al., Delay optimization using SOP balancing, in *Proceedings of the ICCAD* (2011)
45. J.P. Fishburn, A depth-decreasing heuristic for combinational logic, in *Proceedings of the DAC* (1990)
46. D. Nikonov, I. Young, Benchmarking of beyond-CMOS exploratory devices for logic integrated circuits. IEEE J. Explor. Solid-State Comput. Devices Circuits **1**(99), 3–11 (2015)
47. J. Kim et al., Spin-based computing: device concepts, current status, and a case study on a high-performance microprocessor. Proc. IEEE **103**(1), 106–130 (2015)
48. O. Zografos, et al, System-level assessment and area evaluation of spin wave logic circuits, in *IEEE/ACM International Symposium on Nanoscale Architectures (NANOARCH)*. IEEE (2014)
49. D.E. Nikonov et al., Overview of beyond-CMOS devices and a uniform methodology for their benchmarking. Proc. IEEE **101**(12), 2498–2533 (2013)
50. A. Khitun, K.L. Wang, Nano scale computational architectures with spin wave bus. Superlattices Microstruct. **38**(3), 184–200 (2005)
51. A. Khitun et al., Non-volatile magnonic logic circuits engineering. J. Appl. Phys. **110**, 034306 (2011)
52. O. Zografos et al., Majority logic synthesis for spin wave technology, in *17th Euromicro Conference on Digital System Design (DSD)*. IEEE (2014)
53. G.W. Burr et al., Overview of candidate device technologies for storage-class-memory. IBM J. Res. Dev. **52**(4/5), 449–464 (2008)
54. R. Fackenthal et al., A 16Gb ReRAM with 200MB/s write and 1GB/s read in 27nm technology, ISSCC Tech. Dig. (2014)
55. S.-S. Sheu et al., A 4Mb embedded SLC resistive-RAM macro with 7.2ns read-write random-access time and 160ns MLC-access capability, ISSCC Tech. Dig. (2011)
56. H.-S.P. Wong et al., Metal-oxide RRAM. Proc. IEEE **100**(6), 1951–1970 (2012)
57. E. Linn, R. Rosezin, C. Kügeler, R. Waser, Complementary resistive switches for passive nanocrossbar memories. Nat. Mater. **9**, 403–406 (2010)

58. E. Linn, R. Rosezin, S. Tappertzhofen, U. Böttger, R. Waser, Beyond von Neumann–logic operations in passive crossbar arrays alongside memory operations. Nanotechnology **23**, 305205 (2012)
59. P.-E. Gaillardon, L. Amaru, A. Siemon, E. Linn, A. Chattopadhyay, G. De Micheli, Computing secrets on a resistive memory array, (WIP poster) Design Automation Conference (DAC), San Francisco, CA, USA (2015)
60. A. Bogdanov et al., PRESENT: an ultra-lightweight block cipher, CHES Tech. Dig. (2007)
61. E. Sentovich et al., SIS: a system for sequential circuit synthesis, ERL, Department of EECS, University of California, Berkeley, UCB/ERL M92/41 (1992)
62. O. Zografos et al., Majority logic synthesis for spin wave technology, in *Proceedings of the DSD'14*
63. P.D. Tougaw, C.S. Lent, Logical devices implemented using quantum cellular automata. J. Appl. Phys. **75**(3), 1811–1817 (1994)
64. W. Li et al., Three-input majority logic gate and multiple input logic circuit based on DNA strand displacement. Nano Lett. **13**(6), 2980–2988 (2013)
65. P.-E. Gaillardon et al., Computing secrets on a resistive memory array, in *Proceedings of the DAC'15*
66. L. Amaru et al., Efficient arithmetic logic gates using double-gate silicon nanowire FETs, in *Proceedings of the NEWCAS* (2013)
67. K. Bernstein et al., Device and architecture outlook for beyond CMOS switches. Proc. IEEE **98**(12), 2169–2184 (2010)
68. S. Miryala et al., Exploiting the expressive power of graphene reconfigurable gates via post-synthesis optimization, in *Proceedings of the GLVSLI'15*
69. R. Fackenthal et al., A 16Gb ReRAM with 200MB/s write and 1GB/s read in 27nm technology, ISSCC Tech. Dig. (2014)
70. S.-S. Sheu et al., A 4Mb embedded SLC resistive-RAM macro with 7.2ns read-write random-access time and 160ns MLC-access capability, ISSCC Tech. Dig. (2011)

Part II
Logic Satisfiability and Equivalence Checking

The second part of this book is dedicated to formal verification methods. It deals with two main topics: logic satisfiability and equivalence checking.

For logic satisfiability, a nontrivial circuit duality between tautology and contradiction check is introduced, which can speed up SAT tools. Also, an alternative Boolean satisfiability framework based on majority logic is proposed. For equivalence checking, a new approach to verify faster the combinational equivalence between two reversible logic circuits is presented.

Chapter 4
Exploiting Logic Properties to Speedup SAT

In this chapter, we establish a *non-trivial* duality between tautology and contradiction check to speed up circuit SAT. Tautology check determines if a logic circuit is *true* in every possible interpretation. Analogously, contradiction check determines if a logic circuit is *false* in every possible interpretation. A *trivial* transformation of a (tautology, contradiction) check problem into a (contradiction, tautology) check problem is the *inversion* of all outputs in a logic circuit. In this work, we show that *exact* logic *inversion* is not necessary. We give operator switching rules that selectively exchange tautologies with contradictions, and viceversa. Our approach collapses into logic *inversion* just for tautology and contradiction extreme points but generates *non-complementary* logic circuits in the other cases. This property enables computing benefits when an alternative, but equisolvable, instance of a problem is easier to solve than the original one. As a case study, we investigate the impact on SAT. There, our methodology generates a dual SAT instance solvable in parallel with the original one. This concept can be used on top of any other SAT approach and does not impose much overhead, except having to run two solvers instead of one, which is typically not a problem because multi-cores are wide-spread and computing resources are inexpensive. Experimental results show a 25 % speed-up of SAT in a concurrent execution scenario. Also, statistical experiments confirmed that our runtime reduction is not of the random variation type.

4.1 Introduction

Inspecting the properties of logic circuits is pivotal to logic applications for computers and especially to *Electronic Design Automation* (EDA) [1]. There exists a large variety of properties to be checked in logic circuits, e.g., *unateness*, *linearity*, *symmetry*, *balancedness*, *monotonicity*, *thresholdness* and many others [2]. Basic characteristics are usually verified first to provide grounds for more involved tests. Tautology and contradiction are the most fundamental properties in logic circuits. A check for tautology determines if a logic circuit is *true* for all possible input patterns. Analogously, a

© Springer International Publishing Switzerland 2017
L.G. Amaru, *New Data Structures and Algorithms for Logic Synthesis and Verification*, DOI 10.1007/978-3-319-43174-1_4

check for contradiction determines if a logic circuit is *false* for all possible input patterns. While investigating elementary properties, tautology and contradiction check are difficult problems, i.e., co-NP-complete and NP-complete, respectively [3]. Indeed, both tautology and contradiction check are equivalent formulation of the Boolean *SATisfiability* (SAT) problem [3]. In this scenario, new efficient algorithms for tautology/contradiction check are key to push further the edge of computational limits, enabling larger logic circuits to be examined.

Tautology and contradiction check are dual problems. One can interchangeably check for tautology in place of contradiction by *inverting* all outputs in a logic circuit. In this *trivial* approach, the two obtained problems are fully complementary and there is no explicit computational advantage in solving one problem instead of the other.

In this chapter, we show that *exact logic inversion* is not necessary for transforming tautology into contradiction, and *viceversa*. We give a set of operator switching rules that selectively exchange tautologies with contradictions. A logic circuit modified by our rules is *inverted* just if identically *true* or *false* for all input combinations. In the other cases, it is not necessarily the complement of the original one. For this reason, our approach is different from traditional *DeMorganization*. In a simple logic circuit made of AND, OR and INV logic operators, our switching rules swap AND/OR operator types. We give a set of rules for general logic circuits in the rest of this chapter. Note that in this chapter we mostly deal with single output circuits. For multi-output circuits, the same approach can be extended by ORing (contradiction) or ANDing (tautology) the outputs that need to be checked into a single one.

Our approach generates two different, but equisolvable, instances of the same problem. In this scenario, solving both of them in parallel enables a positive computation speed-up. Indeed, the instance solved first stops the other reducing the runtime. This concept can be used on top of any other checking approach and does not impose much overhead, except having to run two solvers instead of one, which is typically not a problem because multi-cores are wide-spread and computing resources are inexpensive. Note that other pallel checking techniques exist. For example, one can launch in parallel many randomized check runs on the same problem instance with the aim to hit the instance-intrinsic minimum runtime [4]. Instead, in our methodology, we create a different but equi-checkable instance that has a potentially lower minimum runtime. As a case study, we investigate the impact of our approach on SAT. There, by using *non-trivial* and *trivial* dualities in sequence, we create a dual SAT instance solvable in parallel with the original one. Experimental results show 25 % speed-up of SAT, on average, in a concurrent execution scenario. Also, statistical experiments confirmed that our runtime reduction is not of the random variation type.

The remainder of this chapter is organized as follows. Section 4.2 describes some background and discusses the motivation for this study. Section 4.3 presents theoretical results useful for the scope of this paper. Section 4.4 proves our main result on the duality between tautology and contradiction check. Section 4.5 shows the benefits enabled by this duality in SAT solving. Section 4.6 concludes the chapter.

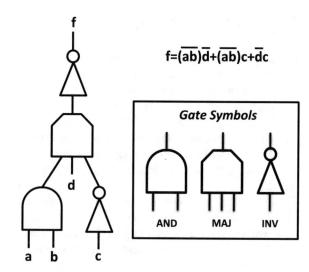

Fig. 4.1 Logic circuit
example representing the
function
$f = \overline{(ab)d} + \overline{(ab)}c + \overline{d}c.$
The *basis* set is {AND, MAJ,
INV}. The gates symbolic
representation is shown in
the box

4.2 Background and Motivation

This section first provides notation on logic circuits. Then, it gives a brief background
on tautology checking from an EDA perspective. Finally, it discusses the motivation
for this study.

4.2.1 Notation

Similarly to the notation used in Chaps. 2 and 3, a logic circuit is modeled by a
Directed Acyclic Graph (DAG) representing a Boolean function, with nodes corresponding to logic gates and directed edges corresponding to wires connecting the
gates. The *on-set* of a logic circuit is the set of input patterns evaluating to *true*. Analogously, the *off-set* of a logic circuit is the set of input patterns evaluating to *false*.
Each logic gate is associated with a primitive Boolean function taken from a prede-
fined set of *basis* logic operators, e.g., AND, OR, XOR, XNOR, INV, MAJ, MIN etc.
Logic operators such as MAJ and MIN represent *self dual* Boolean functions, i.e.,
functions whose output complementation is equivalent to inputs complementation.
A set of *basis* logic operators is said to be *universal*[1] if any Boolean function can
be represented by a logic circuit equipped with those logic gates. For example, the
basis set {OR, INV} is *universal* while the *basis* set {AND, MAJ} is not.

Figure 4.1 shows a logic circuit for the function $f = \overline{(ab)d} + \overline{(ab)}c + \overline{d}c$ over
the *universal basis* set {AND, MAJ, INV}.

---

[1] In this chapter, the term *basis* does not share the same properties as in linear algebra. In particular,
here not all the *basis* are *universal*.

4.2.2 Tautology Checking

Tautology checking, i.e., verifying whether a logic circuit is *true* in every possible interpretation, is an important task in computer science and at the core of EDA [5, 7]. Traditionally, tautology checking supports digital design verification through combinational equivalence checking [7]. Indeed, the equivalence between two logic circuits can be detected by XNOR-ing and checking for tautology. Logic synthesis also uses tautology checking to (i) highlight logic simplifications during optimization [5, 6] and to (ii) identify matching during technology mapping [8]. On a general basis, many EDA tasks requiring automated deduction are solved by tautology check routines.

Unfortunately, solving a tautology check problem can be a difficult task. In its most general formulation, the tautology check problem is co-NP-complete. A straightforward method to detect a tautology is the exhaustive exploration of a function truth table. This *naive* approach can declare a tautology only in exponential runtime. More intelligent methods have been developed in the past. Techniques based on cofactoring trees and binary recursion have been presented in [9]. Together with rules for pruning/simplifying the recursion step, these techniques reduced the checking runtime on several benchmarks. Another method, originally targeting propositional formulas, is Stalmarck's method [10] that rewrites a formula with a possibly smaller number of connectives. The derived equivalent formula is represented by triplets that are propagated to check for tautology. Unate recursive cofactoring trees and Stalmarck's method are as bad as any other tautology check method in the worst case but very efficient in real-life applications. With the rise of *Binary Decision Diagrams* (BDDs) [11], tautology check algorithms found an efficient canonical data structure explicitly showing the logic feature under investigation [12]. The BDD for a tautology is always a single node standing for the logic constant *true*. Hence, it is sufficient to build a BDD for a logic circuit and verify the resulting graph size (plus the output polarity) to solve a tautology check problem. Unfortunately, BDDs can be exponential in size for some functions (multipliers, hidden-weight bit, etc.). In the recent years, the advancements in SAT solving tools [13, 14] enabled more scalable approaches for tautology checking. Using the *trivial* duality between tautology and contradiction, SAT solvers can be used to determine if an *inverted* logic circuit is unsatisfiable (contradiction) and consequently if the original circuit is a tautology. Still, SAT solving is an NP-complete problem so checking for tautology with SAT is difficult in general.

4.2.3 Motivation

Tautology checking is a task surfing the edge of today's computing capabilities. Due to its co-NP-completeness, tautology checking aggressively consumes computational power when the size of the problem increases. To push further the boundary of examinable logic circuits, it is important to study new efficient checking methodologies.

Indeed, even a narrow theoretical improvement can generate a speed-up equivalent to several years of technology evolution.

In this chapter, we present a *non-trivial* duality between contradiction and tautology check problems that opens up new efficient solving opportunities.

4.3 Properties of Logic Circuits

In this section, we show properties of logic circuits with regard to their *on-set/off-set* balance and distribution. These theoretical results will serve as grounds for proving our main claim in the next section.

We initially focus on two *universal basis* sets: {AND, OR, INV} and {MAJ, INV}. We deal with richer *basis* sets later on. We first recall a known fact about majority operators.

Property A MAJ operator of n-variables, with n odd, can be configured as an $\lceil n/2 \rceil$-variables AND operator by biasing $\lfloor n/2 \rfloor$ inputs to logic *false* and can be configured as an $\lceil n/2 \rceil$-variables OR operator by biasing $\lfloor n/2 \rfloor$ inputs to logic *true*.

For the sake of clarity, an example of a three-input MAJ configuration in AND/OR is depicted by Fig. 4.2. Extended at the circuit level, such property enables the emulation of any {AND, OR, INV} logic circuit by a structurally identical {MAJ, INV} logic circuit. This result was previously shown in [2] where logic circuit over the *basis* set {AND, OR, INV} are called AND/OR-INV graphs and logic circuits over the *basis* set {MAJ, INV} are called MAJ-INV graphs.

An example of two structurally, and functionally, identical logic circuits over the *basis* sets {AND, OR, INV} and {MAJ, INV} is depicted by Fig. 4.3a, b. The Boolean function represented in this example is $f = ab + ac + a(b + c) + \bar{a}$. MAJ are configured to behave as AND/OR by fixing one input to *false*(F)/*true*(T), respectively. In place of biasing one input of the MAJ with a logic constant, it is also possible to introduce a fictitious input variable connected in regular/inverted polarity to substitute *true*(T)/*false*(F) constants, respectively. In this way, the function represented is changed but still including the original one when the fictitious input variable is assigned to *true*. Figure 4.3d shows a logic circuit with a fictitious input variable d replacing the logic constants in Fig. 4.3b. The Boolean function represented there is h with property $h_{d=true} = f$.

Fig. 4.2 AND/OR configuration of a three-input MAJ

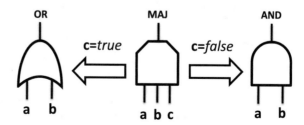

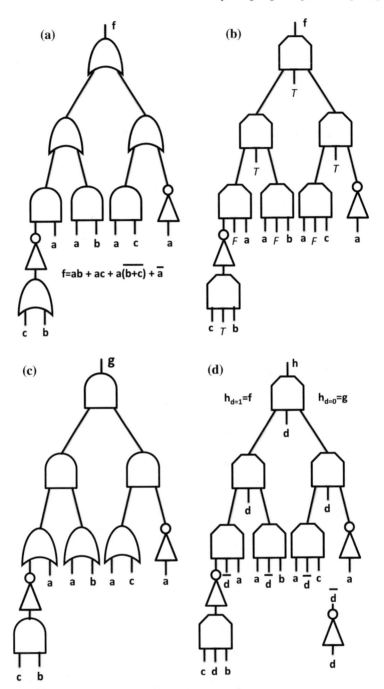

Fig. 4.3 Logic circuits examples. {AND, OR, INV} logic circuit representing $f = ab + ac + a\overline{(b + c)} + \overline{a}$ **a**. {MAJ, INV} logic circuit emulating the circuit in **a** using constants **b**. {AND, OR, INV} logic circuits derived from **a** by switching AND/OR operators **c**. {MAJ, INV} logic circuit emulating the circuit in **a** using an fictitious input variable d **d**

Up to this point, we showed that {AND, OR, INV} logic circuits can be emulated by {MAJ, INV} logic circuits configured either by (i) logic constants or by (ii) a fictitious input variable. In the latter case, {MAJ, INV} logic circuits have all inputs assignable. With no logic constants appearing and all operators being *self-dual*, this particular class of logic circuits have a perfectly balanced *on-set/off-set* size. The following theorem formalizes this property.

Theorem 4.1 *Logic circuits over the* universal basis *set {MAJ, INV}, with all inputs assignable (no logic constants), have* $|\text{on-set}|=2^{n-1}$ *and* $|\text{off-set}|=2^{n-1}$, *with n being the number of input variables.*

Proof MAJ and INV logic operators, with no constants, represent *self-dual* Boolean functions. In [5], it is shown that *self-dual* Boolean functions have an $|on\text{-}set|=|off\text{-}set|=2^{n-1}$, with n being the number of input variables. Also, it is shown in [5] that Boolean functions composed by *self-dual* Boolean functions are *self-dual* as well. This is indeed the case for {MAJ, INV} logic circuits with no constants in input. As these circuits represent *self-dual* Boolean functions, we can assert $|on\text{-}set|=|off\text{-}set|=2^{n-1}$. ∎

{MAJ, INV} logic circuits with no constants have a perfectly balanced partition between *on-set* size and *off-set* size. This is the case for the example in Fig. 4.3d. Eventually, we know that by assigning d to *true* in such example circuit the *on-set/off-set* balance can be lost. Indeed, with $d=true$ the {MAJ, INV} logic circuit then emulates the original {AND, OR, INV} logic circuit in Fig. 4.3a, that could have different *on-set* size and *off-set* size. Still, it is possible to reclaim the perfect *on-set/off-set* balance by superposing the cases $d = true$ and $d = false$ in the {MAJ, INV} logic circuit. While we know precisely what the {MAJ, INV} logic circuit does when $d = true$, the case $d = false$ is not as evident. We can interpret the case $d = false$ as an inversion in the MAJ configuration polarity. This means that where a MAJ is configured as an AND (OR) node in $d = true$, it is instead configured as an OR (AND) node in $d = false$. In other words, $d = false$ in the {MAJ, INV} logic circuit of Fig. 4.3d corresponds to switch AND/OR operator types in the original {AND, OR, INV} logic circuit of Fig. 4.3a. The resulting AND/OR switched circuit is depicted by Fig. 4.3c.

United by a common {MAJ, INV} generalization, {AND, OR, INV} logic circuits and their AND/OR switched versions share strong properties about *on-set/off-set* repartition. The following theorem states their relation.

Theorem 4.2 *Let A be a logic circuit over the* universal basis *set {AND, OR, INV}. Let A' be a modified version of A, with AND/OR operators switched. The following identities hold* $|\text{on-set}(A)| = |\text{off-set}(A')|$ *and* $|\text{off-set}(A)| = |\text{on-set}(A')|$.

Proof Say M a {MAJ, INV} logic circuit emulating A using an extra fictitious input variable, say d. $M_{d=1}$ is structurally and functionally equivalent to A, while $M_{d=0}$ is structurally and functionally equivalent to A'. From Theorem 4.1 we know that $|on\text{-}set(M)| = |off\text{-}set(M)| = 2^{n-1} = 2^m$, where m is the number of input variables in A and

n the number of input variables in M, with $n = m + 1$ to take into account the extra fictitious input variable in M. We know by construction that $|on\text{-}set(M_{d=1})|+|on\text{-}set(M_{d=0})| = 2^{n-1} = 2^m$ and $|off\text{-}set(M_{d=1})|+|off\text{-}set(M_{d=0})| = 2^{n-1} = 2^m$. Again by construction we know that $M_{d=1}$ and $M_{d=0}$ can be substituted by A and A', respectively, in all equations. Owing to the basic definition of A and A' we have that $|on\text{-}set(A)|+|off\text{-}set(A)| = 2^m$ and $|on\text{-}set(A')|+|off\text{-}set(A')| = 2^m$. Expressing $|on\text{-}set(A)|$ as $2^m - |on\text{-}set(A')|$ from the first set of equations and substituting this term in $|on\text{-}set(A)|+|off\text{-}set(A)| = 2^m$ we get $2^m - |on\text{-}set(A')|+|off\text{-}set(A)| = 2^m$ that can be simplified as $|off\text{-}set(A)| = |on\text{-}set(A')|$. This proves the first identity of the Theorem. The second identity can be proved analogously.   ∎

Informally, the previous theorem says that by switching AND/OR operators in an {AND, OR, INV} logic circuit we swap the *on-set* and *off-set* sizes. From a statistical perspective, this is equivalent to invert $Pr(A = true)$ with $Pr(A = false)$, under uniformly random input string of bits. While this also happens with *exact* logic *inversion*, here the actual distribution of the *on-set/off-set* elements is not necessarily complementary. In the next section, we show the implications of the theoretical results seen so far in tautology and contradiction check problems.

4.4 From Tautology to Contradiction and Back

Verifying whether a logic circuit is a tautology, a contradiction or a contingency[2] is an important task in logic applications for computers.

In this section, we show that tautology and contradiction check in logic circuits are dual and interchangeable problems that do not require *exact* logic *inversion per se*. We start by considering logic circuit over the *universal basis* set {AND, OR, INV} and we consider richer *basis* sets later on. The following theorem describes the *non-trivial* duality between tautology and contradiction in {AND, OR, INV} logic circuits.

Theorem 4.3 *Let A be a logic circuit over the* universal basis *set {AND, OR, INV} representing a tautology (contradiction). The logic circuit A', obtained by switching AND/OR operations in A, represents a contradiction (tautology).*

Proof If A represents a tautology then $|on\text{-}set(A)| = 2^m$ and $|off\text{-}set(A)| = 0$, with m being the number of inputs. Owing to Theorem 4.2 $|on\text{-}set(A')| = |off\text{-}set(A)| = 0$ and $|off\text{-}set(A')| = |on\text{-}set(A)| = 2^m$. It follows that A' is a contradiction. Analogous reasoning holds for contradiction to tautology transformation.   ∎

Switching AND/ORs in an {AND, OR, INV} logic circuit is strictly equivalent to logic inversion only for tautology and contradiction. In the other cases, A and A' are not necessarily complementary. We give empirical evidences about this fact

---

[2]A logic circuit is a contigency when it is neither a tautology nor a contradiction [5].

Fig. 4.4 Comparison between *real inverted* and *AND/OR switched* logic circuits representing 4-variable Boolean functions. The *on-set* size ranges from 0 to 2^4

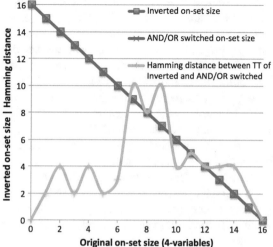

4-Variables AND/OR-switched vs. Real Inverted Logic Circuits

- ■ Inverted on-set size
- ✳ AND/OR switched on-set size
- ● Hamming distance between TT of Inverted and AND/OR switched

Inverted on-set size | Hamming distance (y-axis)

Original on-set size (4-variables) (x-axis)

hereafter. Figure 4.4 depicts the obtained results in a graph chart. We examined 17 random Boolean functions of four input variables, with *on-set size* ranging from 0 (contradiction) to 16 (tautology). We first compared the *on-set* size of the real inverted logic circuits with the *on-set size* of the AND/OR switched circuits. As expected, Theorem 4.2 holds and switching AND/OR operators results in exchanging the *on-set* and *off-set* sizes. This also happens with the real inverted circuits, but in that case also the actual *on-set/off-set* elements distribution is complementary. To verify what is the *on-set/off-set* elements distribution in general, we define a distance metric between the real inverted and AND/OR switched circuits. The distance metric is computed in two steps. First, the truth tables of the circuits are unrolled, using the same input order, and represented as binary strings. Second, the distance metric is measured as the Hamming distance[3] between those binary strings. For tautology and contradiction extremes the distance metric between AND/OR switched circuits and real inverted circuits is 0, as obvious consequence of Theorem 4.3. For other circuits, real inverted and AND/OR switched circuits are different, with distance metric ranging between 2 and 10.

As a practical interpretation of the matter discussed so far, we can get an answer for a tautology (contradiction) check problem by working on a functionally different and *non-complementary* structure than the original one under test. We explain hereafter why this fact is interesting. Suppose that the logic circuit we want to check is a contigency but algorithms for tautology (contradiction) are not efficient on it. If we just invert the outputs of this logic circuit and we run algorithms for contradiction (tautology) then we would likely face the same difficulty. However, if we switch

---

[3]The Hamming distance between two binary strings, of equal size, is the number of positions at which the corresponding bits are different.

AND/ORs in the logic circuit we get a functionally different and *non-complementary* structure. In this case, algorithms for contradiction (tautology) do not face by construction the same complexity. Exploiting this property, it is possible to speed-up a traditional tautology (contradiction) check problem. Still, Theorem 4.3 gurantees that if the original circuit is a tautology (contradiction) then the AND/OR switched version is a contradiction (tautology) preserving the checking correctness.

Recalling the example in Fig. 4.3a, the original logic circuit represents a tautology. Consequently, the logic circuit in Fig. 4.3c represents a contradiction. These properties are verifiable by hand as the circuits considered are small. For an example which is a *contingency*, consider the {AND, OR, INV} circuit realization for $f = ab' + c'$ (contingency). By switching AND/ORs, we get $g = (a + b')c'$ which is different from both f or f', as predicted.

We now consider logic circuits with richer *basis* set functions than just {AND, OR, INV}. Our enlarged *basis* set includes {AND, OR, INV, MAJ, XOR, XNOR} logic operators. Other operators can always be decomposed into this *universal basis* set, or new switching rules can be derived. In the following, we extend the applicability of Theorem 4.3.

Theorem 4.4 *Let A be a logic circuit over the* universal basis *set {AND, OR, INV, MAJ, XOR, XNOR} representing a tautology (contradiction). The logic circuit A', obtained by switching logic operators in A as per Table 4.1, represents a contradiction (tautology).*

Proof In order to prove the theorem, we need to show the switching rules just for XOR, XNOR and MAJ operators. AND/OR switching is already proved by Theorem 4.3. Consider the XOR operator decomposed in terms of {AND, OR, INV}: $f = a \oplus b = ab' + a'b$. Applying the duality in Theorem 4.3 we get $g = (a + b')(a' + b)$ that is indeed equivalent to a XNOR operator. This proves the XOR to XNOR switching and *viceversa*. Analogously, consider the MAJ operator decomposed in terms of {AND, OR, INV}: $f = ab + ac + bc$. Applying the duality in Theorem 4.3 we get $g = (a + b)(a + c)(b + c)$ that is still equivalent to a MAJ operator. Hence, MAJ operators do not need to be modified. ∎

Note that in a data structure for a computer program, the operator switching task does not require actual pre-processing of the logic circuit. Indeed, each time that a node in the DAG is evaluated an external flag word determines if the regular or switched operator type has to be retrieved from memory.

In the current subsection, we showed a *non-trivial* duality between contradiction and tautology check. In the next subsection, we study its application on Boolean satisfiability.

4.4.1 Boolean SAT and Tautology/Contradiction Duality

The Boolean SAT problem consists of determining whether there exists or not an interpretation evaluating to *true* a Boolean formula or circuit. The Boolean SAT

Table 4.1 Switching rules for tautology/contradiction check

Original logic operator	Switched logic operator
INV	INV
AND	OR
OR	AND
MAJ	MAJ
XOR	XNOR
XNOR	XOR

problem is reciprocal to a check for contradiction. When contradiction check fails then Boolean SAT succeeds while when contradiction check succeeds then Boolean SAT fails. Instead of checking for Boolean SAT or for contradiction, one can use a dual transformation in the circuit and check for tautology. Such transformation can be either (i) *non-trivial*, i.e., switching logic operators in the circuit as per Table 4.1 or (ii) *trivial*, i.e., output complementation. If we use twice any dual transformation, we go back to the original problem domain (contradiction, SAT). Note that if we use twice the same dual transformation (*trivial-trival* or *non-trivial-non-trival*) we obtain back exactly the original circuit. Instead, if we apply two different dual transformations in sequence (*trivial-non-trival* or *non-trivial-trival*) we obtain an equisatisfiable but not necessarily equivalent circuit. We use the latter approach to generate a second equisatisfiable circuit, which we call the dual circuit. The pseudocode in Algorithm 6 shows our speculative SAT flow.

Algorithm 6 Speculative parallel regular/dual circuit SAT pseudocode.

INPUT: Logic circuit α.　　　　**OUTPUT:** SAT/unSAT solution for α.

α'=Dual(α);// non-trivial duality from Table 4.1 - can be done while reading α
α'=NOT(α');// output complementation - can be done while reading α
solution =∅;
while solution is ∅ **do**
　　solution=SAT(α) ‖ solution=SAT(α');// solve in parallel the SAT problem for α and α', the first finishing stops the execution
end while
return solution;

First, the dual circuit is built by first applying our *non-trivial duality* (switching rules in Table 4.1). Then, the dual circuit is modified by complementing the outputs (*trivial* duality). Note that these two operations can be done while reading the regular circuit itself, thus ideally require no (or very little) computational overhead, as explained previously. Finally, the dual circuit SAT is solved in parallel with the regular one in a *"first finishing wins"* speculative strategy. Figure 4.5 graphically depicts the flow.

Fig. 4.5 Speculative parallel regular/dual circuit SAT flow

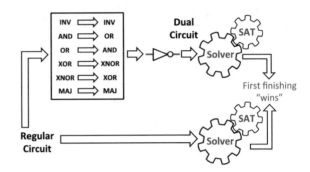

4.5   Experimental Results

In this section, we exercise our *non-trivial* duality in Boolean *SATisfiability* (SAT) problems. First, we demonstrate that the dual instance can be solved faster than the regular one and the corresponding runtime reduction is not of the random variation type. Second, we show experimental results for a concurrent regular/dual SAT execution scenario.

4.5.1   *Verification of SAT Solving Advantage on the Dual Circuit*

In our first set of experiments we focused on verifying whether the dual circuit can be easier to satisfy than the regular circuit. For this purpose, we modified MiniSat-C v1.14.1 [16] to read circuits in AIGER format [18] and to encode them in CNF internally via Tseitin transformation. The dual circuit is generated online during reading if a switch "-p" is given. We considered a large circuit (0.7 M nodes) and we created 1000 randomized SAT instances by setting the number generator seed in MiniSat to *rand()*.   The plot in Fig. 4.6 shows the number of instances (Y axis) solved in a given execution time (X axis), for both the dual and regular SAT flows. More specifically, the two curves Fig. 4.6 represent the runtime distributions for dual and regular SAT. The dual runtime distribution is clearly left-shifted (but partially overlapping) with respect to the regular runtime distribution. This confirms that (i) the dual circuit can be solved faster than the regular one and (ii) the runtime reduction is not of the random variation type.

4.5.2   *Results for Concurrent Regular/Dual SAT Execution*

In our second set of experiments (downloadable at [19]) we used ABC tool [17] to test our dual approach together with advanced techniques to speed-up SAT. Our custom set of benchmarks is derived by (i) unfolding SAT sequential problems (ii) encoding

Fig. 4.6 1000 randomized
SAT runs for regular and
dual circuit

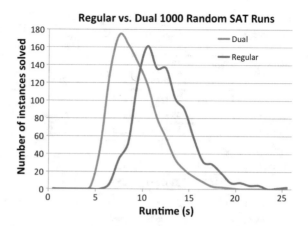

combinational equivalence check problems. All benchmarks are initially described
in Verilog as a netlist of logic gates over the *basis* {AND, OR, INV, XOR, XNOR,
MAJ}. The dual circuits are obtained by applying switching rules in Table 4.1 and
inverting the output. The ABC script to read and run SAT on these benchmarks is:
*read library.genlib; r -m input.v; st; write out.aig; &r out.aig; &ps; &write_cnf -K 4
out.cnf; dsat -p out.cnf.* Apart from standard I/O commands, note that *&write_cnf -K
4 out.cnf* generates a CNF using a technology mapping procedure and *dsat -p* calls
MiniSat with variable polarity alignment.

Table 6.1 shows results for regular versus dual SAT solving with our setup
(Table 4.2). For about half of the benchmarks (7/13) the dual instance concluded
first while for the remaining ones (6/13) the regular instance was faster. The total
regular runtime is quite close to the total dual runtime (just 6% of deviation). However, considering here the speculative parallel SAT flow in Fig. 4.5, we can ideally
reduce the total runtime by about 25%. Note that this is an ideal projection into a
parallel execution environment, with no overhead. We experimentally verified that
the average overhead can be small (few percentage points) thanks to the intrinsic
independence of the two tasks.

4.6 Summary

In this chapter, we presented a *non-trivial* duality between tautology and contradiction check to speed up circuit SAT. On the one hand, tautology check determines if a logic circuit is *true* for all input combinations. On the other hand,
contradiction check determines if a logic circuit is *false* for all input combinations.
A *trivial* transformation of a (tautology, contradiction) check problem into a (contradiction, tautology) check problem is the *inversion* of all the outputs in a logic circuit.
In this work, we proved that *exact* logic *inversion* is not necessary. By switching
logic operator types in a logic circuit, following the rules presented in this paper, we

Table 4.2 Experimental results for regular versus dual SAT solving all runtimes are in seconds

Benchmark	I/O	Logic size	Logic depth	Runtime regular	Runtime dual	$|\Delta$ runtime$|$	Best runtime
hardsat1	4580/1	283539	392	186.35	58.9	127.35	58.9
hardsat2	4580/1	287635	392	51.1	191.87	140.77	51.1
hardsat3	198540/1	920927	267	0.94	1.1	0.16	0.94
hardsat4	2452/1	43962	436	68.82	20.53	48.29	20.53
hardsat5	5725/1	562027	464	40.91	22.72	18.19	22.72
hardsat6	3065/1	86085	437	37.51	64.24	26.73	37.51
hardsat7	372240/1	85596	151	4.8	3.68	1.12	3.68
Total *sat*	591182/7	2269771	2539	390.43	363.04	27.39	195.38
hardunsat1	61/1	448884	2181	26.72	27.22	0.50	26.72
hardunsat2	61/1	264263	2951	3.70	1.32	2.38	1.32
hardunsat3	61/1	451350	2181	27.8	20.33	7.47	20.33
hardunsat4	540/1	244660	1158	234.88	326.84	91.96	234.88
hardunsat5	2352/1	208221	439	7.61	7.65	0.04	7.65
hardunsat6	550/1	117820	423	142.28	137.94	4.34	137.94
Total *unsat*	3625/6	1735198	9333	442.99	521.30	78.31	428.80
Total	594807/13	4004969	11872	833.42	884.34	50.84	624.18
Norm. to Regular	–	–	–	**1.00**	1.06	–	**0.75**

can selectively exchange tautologies with contradictions. Our approach is equivalent to logic *inversion* just for tautology and contradiction extreme points. It generates *non-complementary* logic circuits in the other cases. Such property enables computing benefits when an alternative but equisolvable instance is easier to solve than the original one. As a case study, we studied the impact on SAT. There, our methodology generated a dual SAT instance solvable in parallel with the original one. This concept can be used on top of any other SAT approach and does not impose much overhead, except having to run two solvers instead of one, which is typically not a problem because multi-cores are wide-spread and computing resources are inexpensive. Experimental results shown 25 % speed-up of SAT in a concurrent execution scenario.

References

1. G. De Micheli, *Synthesis and Optimization of Digital Circuits* (McGraw-Hill, New York, 1994)
2. T. Sasao, *Switching Theory for Logic Synthesis* (Springer, Heidelberg, 1999)
3. M.R. Garey, D.S. Johnson, *Computers and Intractability- A Guide to the Theory of NP-Completeness* (W. H Freeman and Company, New York, 1979)

4. A.E. Hyvarinen et al., Incorporating clause learning in grid-based randomized SAT solving. J SAT (JSAT) **6**, 223–244 (2009)
5. R.K. Brayton, *Logic Minimization Algorithms For VLSI Synthesis*, vol. 2, (Springer, Heidelberg, 1984)
6. R. Rudell, A. Sangiovanni-Vincentelli, *Multiple-valued Minimization far PLA Optimization*, IEEE Trans. CAD ICs Syst. **6**(5) 727–750 (1987)
7. G. Hachtel, F. Somenzi, *Logic Synthesis And Verification Algorithms* (Springer, Heidelberg, 2006)
8. L. Benini, G. De Micheli, A survey of Boolean matching techniques for library binding. ACM Trans. DAES (TODAES) **2**(3), 193–226 (1997)
9. G.D. Hachtel, M.J. Reily, *Verification Algorithms for VLSI Synthesis*, IEEE Trans. CAD of ICs Syst. **7**(5) 616–640 (1980)
10. G. Stalmarck, *A system for determining propositional logic theorems by applying values and rules to triplets that are generated from a formula*, Swedish Patent No. 467,076 (approved 1992); U.S. Patent No. 5,276,897 (approved 1994); European Patent No. 403,454 (approved 1995)
11. R.E. Bryant, *Graph-based Algorithms For Boolean Function Manipulation*, IEEE Trans. Comp. **C-35**(8) 677–691 (1986)
12. S. Malik, A.R. Wang, R.K. Brayton, A. Sangiovanni-Vincentelli, *Logic Verification Using Binary Decision Diagrams In A Logic Synthesis Environment* (Proc, ICCAD, 1988)
13. C.P. Gomes, H. Kautz, A. Sabharwal, B. Selman, Satisfiability solvers. Handb Knowl Represent **3**, 89–134 (2008)
14. http://www.satcompetition.org
15. L. Amaru, P.-E. Gaillardon, G. De Micheli, *Majority-Inverter Graph: A Novel Data-Structure and Algorithms for Efficient Logic Optimization* (Proc, DAC, 2014)
16. MiniSat SAT solver. http://minisat.se/MiniSat.html
17. Berkeley Logic Synthesis and Verification Group, ABC: A System for Sequential Synthesis and Verification, http://www.eecs.berkeley.edu/~alanmi/abc/
18. AIGER benchmarks available online at http://fmv.jku.at/aiger/
19. http://lsi.epfl.ch/DUALSAT

Chapter 5
Majority Normal Form Representation and Satisfiability

In this chapter, we focus on a novel two-level logic representation. We define *Majority Normal Form* (MNF), as an alternative to the traditional *Disjunctive Normal Form* (DNF) and the *Conjunctive Normal Form* (CNF). After a brief investigation on the MNF expressive power, we study the problem of MNF-*SATisfiability* (MNF-SAT). We prove that MNF-SAT is NP-complete, as its CNF-SAT counterpart. However, we show practical restrictions on MNF formula whose satisfiability can be decided in polynomial time. We finally propose a simple algorithm to solve MNF-SAT, based on the intrinsic functionality of two-level majority logic. Although an automated MNF-SAT solver is still under construction, manual examples already demonstrate promising opportunities.

5.1 Introduction

As shown in the previous chapters of this book, Boolean logic is commonly defined in terms of primitive AND ($\cdot$), OR ($+$) and INV ($'$) operators. Such formulation acts in accordance with the natural way logic designers interpret Boolean functions. For this reason, it emerged as a standard in the field. However, no evidence is provided that this formulation, or another, has the most efficient set of primitives for Boolean logic. In computer science, the efficiency of Boolean logic applications is measured by different metrics such as (i) the result quality, for example the performance of an automatically synthesized digital circuit, (ii) the runtime and (iii) the memory footprint of a software tool. With the aim to optimize these metrics, the accordance to a specific logic model is no longer important. Majority logic has shown the opportunity to enhance the efficiency of multi-level logic optimization [1, 2] and reversible quantum logic synthesis [3].

In this chapter, we extend the intuition provided in Chap. 3 to two-level logic and Boolean satisfiability. We provide an alternative two-level representation of Boolean functions based entirely on majority and complementation operators. We call it *Majority Normal Form* (MNF), using a similar notation as for traditional

© Springer International Publishing Switzerland 2017
L.G. Amaru, *New Data Structures and Algorithms for Logic Synthesis and Verification*, DOI 10.1007/978-3-319-43174-1_5

Disjunctive Normal Form (DNF) and *Conjunctive Normal Form* (CNF) [4]. The MNF can represent any Boolean function, therefore being *universal*, as CNF and DNF. We investigate then the satisfiability of MNF formula (MNF-SAT). In its most general definition, MNF-SAT is NP-complete, as its CNF-SAT counterpart. However, there exist interesting restrictions of MNF whose satisfiability can instead be decided in polynomial time. We finally propose an algorithm to solve MNF-SAT exploiting the nature of two-level majority logic. Manual examples on such algorithm already demonstrate promising opportunities.

The remainder of this chapter is organized as follows. Section 5.2 provides relevant background and notations. In Sect. 5.3, the two-level *Majority Normal Form* is introduced and its features investigated. Section 5.4 studies the satisfiability of MNF formula, from a theoretical perspective. Section 5.5 proposes a simple algorithm to solve MNF-SAT exploiting the intrinsic functionality of two-level majority logic. Section 5.6 discusses future research directions. Section 5.7 concludes the chapter.

5.2 Background and Motivation

This section presents a brief background on two-level logic representation and Boolean satisfiability. Notations and definitions used in the rest of this paper are also introduced.

5.2.1 Notations and Definitions

In the binary Boolean domain, all variables belong to $\mathbb{B} = \{0, 1\}$. The *on*-set of a Boolean function is the set of input patterns evaluating the function to logic 1. Similarly, the *off*-set of a Boolean function is the set of input patterns evaluating the function to logic 0. Literals are variables and complemented ($'$) variables. Terms are conjunctions ($\cdot$) of literals. Clauses are disjunctions ($+$) of literals. A majority function of n (odd) literals returns the Boolean value most frequently appeared among the inputs. In the context of this chapter, we refer to a *threshold* function as to a majority function with repeated literals. Note that this is a restriction of the more general definition of *threshold* functions [5].

5.2.2 Two-Level Logic Representation

Traditional two-level logic representation combines terms and clauses to describe Boolean functions. A *Conjunctive Normal Form* (CNF) is a conjunction of clauses. A *Disjunctive Normal Form* (DNF) is disjunctions of terms. Both CNF and DNF are *universal* logic representation form, i.e., any Boolean function can be represented by them. For more information about logic representation forms, we refer the reader to [5].

5.2.3 Satisfiability

The Boolean *SATisfiability* problem (SAT) has been introduced in Chap. 4. In brief, it consists of determining whether there exists or not an assignment of variables so that a Boolean formula evaluates to true. SAT is a difficult problem for CNF formula. Indeed, CNF-SAT was the first known NP-complete problem [6]. Instead, DNF-SAT is trivial to solve [8]. Unfortunately, converting a CNF into a DNF, or viceversa, may require an exponential number of operations. Some restrictions of CNF-SAT, e.g., 2-SAT, Horn-SAT, XOR-SAT, etc., can be solved in polynomial time. For more information about SAT, we refer to [8].

5.3 Two-Level Majority Representation Form

In this section, we present a two-level majority logic representation form as extension to traditional two-level conjunctive and disjunctive normal forms.

5.3.1 Majority Normal Form Definition and Properties

Both CNF and DNF formula require at least two Boolean operators, $\cdot$ and $+$, apart from the complementation. Interestingly enough, the majority includes both $\cdot$ and $+$ into a unique operator. This feature is formalized in the following.

Property *The n-input (odd) majority operator filled with $\lfloor n/2 \rfloor$ logic zeros collapses into an $\lceil n/2 \rceil$-input $\cdot$ operator. Conversely, if filled with $\lfloor n/2 \rfloor$ logic ones it collapse into an $\lceil n/2 \rceil$-input $+$ operator.*

Example 5.1 Consider the function $M(a, b, c, 0, 0)$. Owing to the majority functionality, to evaluate such function to logic 1 all variables (a, b, c) must be logic 1. This is because already 2 inputs over 5 are fixed to logic 0, which is close to the majority threshold. Indeed, if even only one variable among (a, b, c) is logic 0, the function evaluates to 0. This is equivalent to the function $a \cdot b \cdot c$. Using a similar reasoning, $M(a, b, c, 1, 1)$ is functionally equivalent to $a + b + c$.

This remarkable property motivates us to define a novel two-level logic representation form.

Definition 5.1 A *Majority Normal Form* (MNF) is a majority of majorities, where majorities are fed with literals, 0 or 1.

Example 5.2 An MNF is $M(M(a, b, 1), M(a, b, c, 0, e'), d')$. Another MNF, for a different Boolean function, is $M(a, 0, c, M(a, b', c'), (a', 1, c))$. The expression $M(M(M(a, b, c), d, e), e, f, g, h)$ is not an MNF as it contains three levels of majority operators, while MNF is a two-level representation form.

Following its definition, MNF includes also CNF and DNF.

Property *Any CNF (DNF) is structurally equivalent to an MNF, where the n-input conjunction (disjunction) is a majority operator filled by $\lfloor n/2 \rfloor$ logic zeros (ones) and by n clauses (terms) of m-inputs, that are theirselves majority operators filled by $\lfloor m/2 \rfloor$ logic ones (zeros) and m literals.*

We give hereafter an example of CNF to MNF translation.

Example 5.3 The starting CNF is $(c' + b) \cdot (a' + c) \cdot (a + b)$. The $\cdot$ in the CNF is translated as $M(-, -, -, 0, 0)$. The clauses are instead translated in the form $M(-, -, 1)$. The resulting MNF is $M(M(c', b, 1), M(a', c, 1), M(a, b, 1), 0, 0)$.

It is straightforward now to show that CNF and DNF can be translated into MNF in linear time. However, the inverse translation of MNF into CNF or DNF can be more complex, as MNF are intrinsically more expressive than CNF and DNF.

The MNF is a *universal* logic representation form, i.e., any Boolean function can be represented with it. This comes as a consequence of the inclusion of *universal* CNF and DNF. In addition to the emulation of traditional conjunction and disjunction operators, a majority operator features other noteworthy properties. First, majority is a *self-dual* function [5], i.e., the complement of a majority equals to the majority with complemented inputs. The *self-dual* property also holds when variables are repeated inside the majority operator (*threshold function*). Second, the majority is *fully-symmetric*, i.e., any permutation of inputs does not change the function behavior. In addition, the n-input majority where two inputs are one the complement of the other, collapses into a $(n-2)$-input majority. In order to extend the validity of these properties, it is proper to define $M(a) = a$, which is a majority operator of a single input, equivalent to a logic buffer.

5.3.2 Representation Examples with DNF, CNF and MNF

We provide hereafter some examples of MNF in contrast to their corresponding CNF and DNF.

Example 5.4 Boolean function $a + (b \cdot c)$. The form $a + (b \cdot c)$ is already a DNF. A CNF is $(a + b) \cdot (a + c)$. An MNF is $M(a, 1, M(0, b, c))$. Another, more compact, MNF is $M(a, b, c, a, 1)$.

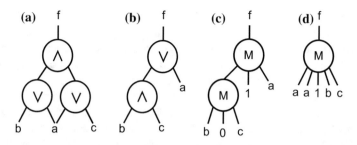

Fig. 5.1 Two-level representation example for the Boolean function $a + (b \cdot c)$ in forms: **a** DNF, **b** CNF, **c** MNF and **d** more compact MNF

For the sake of illustration, Fig. 5.1 depicts the previous example by means of drawings.

Example 5.5 Boolean function $(a \cdot d') + (a \cdot b) + (a \cdot c) + (a' \cdot b \cdot c \cdot d')$. This form is already a DNF. A CNF is $(a + b) \cdot (a + c) \cdot (a + d') \cdot (b + c + d')$. A compact MNF is $M(a, a, b, c, d')$.

Example 5.6 Boolean function $MAJ(a, b, c, d, e)$. A DNF for this function is $(a \cdot b \cdot c) + (a \cdot b \cdot d) + (a \cdot b \cdot e) + (a \cdot c \cdot d) + (a \cdot c \cdot e) + (a \cdot d \cdot e) + (b \cdot c \cdot d) + (b \cdot c \cdot e) + (b \cdot d \cdot e) + (c \cdot d \cdot e)$. As this particular function is monotonic and self-dual, a CNF can be obtained by swapping $\cdot$ and $+$ operators. A compact MNF is simply $M(a, b, c, d, e)$.

Example 5.7 Boolean function $a \oplus b \oplus c$. A DNF is $(a \cdot b \cdot c) + (a \cdot b' \cdot c') + (a' \cdot b \cdot c') + (a' \cdot b' \cdot c)$. A CNF is $(a' + b' + c) \cdot (a' + b + c') \cdot (a + b' + c') \cdot (a + b + c)$ A compact MNF is $M(a, M(a', b, c), M(a', b', c'))$.

Table 5.1 summarizes the sizes of the DNF, CNF and MNF encountered in the previous examples. The size of a CNF size is its number of clauses. Similarly, the size of a DNF is its number of terms. The size of an MNF is the number of majority operators appearing in the formula. As we can notice, the MNF is often more compact than CNF and DNF, with a size ranging from 1 to 4, while the corresponding CNF and DNF sizes range from 2 to 10. Similar results also emerged from theoretical studies

Table 5.1 Two-level logic representation comparison

Boolean Function	DNF Size	CNF Size	MNF Size
$a + (b \cdot c)$	2	2	1
$(a + b) \cdot (a + c) \cdot (a + d') \cdot (b + c + d')$	4	4	1
$(c' + b) \cdot (a' + c) \cdot (a + b)$	3	3	4
$M(a, b, c, d, e)$	10	10	1
$a \oplus b \oplus c$	4	4	3

on circuit complexity [9, 10]. Indeed, it has been shown in [9] that majority circuits of depth 2 and 3 possess the expressive power to represent arithmetic functions, such as powering, multiplication, division, addition etc., in polynomial size. On the other hand, CNF and DNF already require an exponential size for parity, majority and addition functions, which instead are polynomial with MNF [10].

So far, we showed that two-level logic can be expressed in terms of majority operators in place of · and +. This comes at an advantage in representation size as compared to traditional CNF and DNF. Moreover, the natural properties of the majority function permit a uniform and efficient logic manipulation [2]. Still, further investigation and development of the topic are needed, as they will be discussed in Sect. 5.6. In the next section, we study the promising application of MNF formula to Boolean satisfiability.

5.4   Majority Satisfiability

Boolean satisfiability, often abbreviated as SAT, is a core problem of computer science. New approaches to solve SAT, such as [11, 12], are of paramount interest to a wide class of computer applications. This is particularly relevant for *Electronic Design Automation* (EDA).

SAT is in general trivial for some representation form, such as DNF or *Binary Decision Diagrams* (BDDs) [13]. It is instead a difficult problem for CNF formula. For this reason, CNF-SAT is still actively studied. New SAT formulations are of great relevance when their representation can be derived from CNF in polynomial (preferably linear) time. The satisfiability of MNF formula falls in this category as MNF can be derived from CNF in linear time. This fact motivates us to study the general complexity of MNF-SAT.

5.4.1   Complexity of Unrestricted MNF-SAT

To classify the complexity of unrestricted MNF-SAT, we compare it to the well understood CNF-SAT.

Theorem 5.1 *MNF-SAT is NP-complete.*

Proof CNF-SAT is the first known NP-complete problem [6]. Since any CNF formula can be reduced in linear time into a MNF, the complexity of MNF-SAT must also be NP-complete [14].   ∎

Not surprisingly, MNF-SAT is as complex as CNF-SAT. Interestingly enough, alternative proofs, showing that MNF-SAT is a difficult problem, do exist. For example, one can make use of Lewis' representation Theorem [15] or show the reducibility of other NP problems into MNF-SAT [14].

5.4.2 Complexity of Some Restricted MNF-SAT

Even though MNF-SAT is in general a difficult problem, there are restrictions of MNF formula whose satisfiability can be determined easily. We define hereafter some MNF restrictions of interest.

Definition 5.2 MNF_0 is an MNF where logic constant 1 is forbidden (also in the form of $0'$).

Example 5.8 A valid MNF_0 is $M(M(a, b, 0), M(a, b', c), a)$. Instead, $M(M(a, b, 1), c', 0)$ is not an MNF_0 as logic 1 appears inside the formula.

Definition 5.3 MNF_1 is an MNF where logic constant 0 is forbidden (also in the form of $1'$).

Example 5.9 A valid MNF_1 is $M(M(a, 1, d), M(a', b', e), 1)$. Instead, $M(a, 1, M(a', b, 0))$ is not an MNF_1 as logic 0 appears inside the formula.

Definition 5.4 MNF_{pure} is an MNF where both logic constant 1 and logic constant 0 are forbidden.

Example 5.10 A MNF_{pure} is $M(M(a, b, c), M(a, b', c), a')$.

Note that $MNF_0 \supset MNF_{pure}$ and $MNF_1 \supset MNF_{pure}$, but we keep them separated for the sake of reasoning.

Theorem 5.2 MNF_{pure}-SAT is always satisfiable.

Proof In [5], it is proven that a self-dual function fed with other self-dual functions remains self-dual. This is the case for MNF_{pure}, which is indeed always self-dual. A notable property of self-dual functions is to have an on-set of size 2^{n-1}, where n is the number of variables [5]. This means that an MNF_{pure} cannot reach an on-set of size 0 and therefore cannot be unsatisfiable. ∎

Informally, an MNF_1 is an MNF_{pure} with some input biased to logic 1. As MNF_{pure} is always satisfiable as adding more logic 1 to the MNF cannot make it unsatisfiable. Indeed, adding logic ones to an MNF only helps its satisfiability. It follows that also MNF_1 is always satisfiable.

Corollary 5.1 MNF_1-SAT is always satisfiable.

Proof (by contradiction) Without loss of generality, let us assume that an MNF_1 is a fictitious MNF_{pure} where logic 1 is an additional variable, but succesively fixed to 1. Suppose that by moving from the fictitious MNF_{pure} to a MNF_1 we can decrease the *on*-set of size from 2^{n-1} to 0, and therefore make it unsatisfiable. Recall that the majority function is *monotone increasing*, and that *monotonicity* is closed under the composition of functions [5]. By construction, all input vectors to the MNF_1 are bitwise greater or equal as compared to the corresponding input vectors to the fictitious MNF_{pure}. Owing to *monotonicity*, also the MNF_1 evaluates to logic values always greater or equal than the ones of the fictitious MNF_{pure} for the same input vectors. Hence, the *on*-set size of MNF_1 cannot be smaller than the *on*-set size of the fictitious MNF_{pure} and thus cannot reach 0. Here is the contradiction. It follows that MNF_1 formula are always satisfiable.   ∎

The problem of MNF_1-SAT is dual to MNF_0-tautology check.[1] In the following theorem, which is conceptually symmetric to the previous one, we establish their relation.

Theorem 5.3 *MNF_0 is never a tautology.*

Proof (by contradiction) Without loss of generality, let us assume that an MNF_0 is a fictitious MNF_{pure} where logic 0 is an additional variable, but successively fixed to 0. Suppose that by moving from the fictitious MNF_{pure} to a MNF_0 we can increase the *on*-set of size from 2^{n-1} to 2^n, and therefore make it a tautology. Recall that the majority function is *monotone increasing*, and that *monotonicity* is closed under the composition of functions [5]. By construction, all input vectors to the fictitious MNF_{pure} are bitwise greater or equal as compared to the corresponding input vectors to the MNF_0. Owing to *monotonicity*, also the fictitious MNF_{pure} evaluates to logic values always greater or equal than the ones of the MNF_0 for the same input vectors. Hence, the *on*-set size of MNF_0 cannot be greater than the *on*-set size of the fictitious MNF_{pure} and thus cannot reach 2^n. Here is the contradiction. It follows that MNF_0 formula are not tautologies.   ∎

Whenever an MNF can be restricted to MNF_{pure} or MNF_1, its satisfiability is guaranteed, with no need to check. If instead an MNF can be restricted to MNF_0, its tautology check always returns false. We do not focus on algorithms to solve general MNF-SAT or MNF_0-SAT, but we propose in the following section a general methodology applicable to solve MNF-SAT.

5.5 Algorithm to Solve MNF-SAT

In order to automatically solve MNF-SAT instances, an algorithm is needed. We provide a core *decide* algorithm, with linear time complexity with respect to the MNF size. It exploits the intrinsic nature of MNF formula and can be embedded in a traditional *Decide–Deduce–Resolve* SAT solving approach [8]. We start from a one-level majority case and then we move to the two-level MNF case. Note that a recent work [16] considered the satisfiability of two-level (general) threshold circuits. It is proposed to reduce it to a *vector domination problem*. We differentiate from [16] by (i) focusing on MNF formula and (ii) developing a native solving methodology.

---

[1]The tautology check problem has been introduced in Chap. 4 of this book.

5.5.1 One-Level Majority-SAT

In the case of a one-level majority function, the satisfiability check can be accomplished *exactly in linear time* by direct variable assignment (solely *decide* task). Informally, considering a single majority operator, a greedy strategy can maximize the number of logic 1 in an input pattern. If the pattern with the maximum number of logic 1 cannot evaluate a majority to 1, then no other input pattern can do so, because of the majority function *monotonicity*. An automated method for this task is depicted by Algorithm 7 and explained as follows. Each variable is processed in sequence, in any order. If the considered variable appears more often complemented than in its standard polarity, it is set to logic 0, otherwise to logic 1. At the end of this procedure, an assignment for the input variables to the majority operator is obtained. If this assignment cannot evaluate the majority operator to true, then it is declared unsatisfiable, otherwise it is declared satisfiable. An example is provided hereafter.

Algorithm 7 One-level Majority SAT

INPUT: Inputs x_1^n of a majority operator
OUTPUT: Assignment of x_1^n (if SAT this assignment evaluates to true, otherwise unSAT)
  for (i=1; i≤n_vars; i++) **do**
    if x_i appears more often complemented **then**
      $x_i = 0$;
    else
      $x_i = 1$;
    end if
  end for
  if $M(x_1^n)$ evaluates to 1 **then**
    return SAT;
  else
    return unSAT;
  end if

Example 5.11 The Boolean formula whose satisfiability we want to check is $M(a, b, a', a', b, c', c', d, e)$. To find an assignment which evaluates to logic 1, variables are considered in the order (a, b, c, d, e).

Variable a appears more often complemented in the MAJ operator, so it assigned to logic 0.

Variable b appears more often uncomplemented in the MAJ operator, so it assigned to logic 1.

Variable c appears more often complemented in the MAJ operator, so it assigned to logic 0.

Variable d appears more often uncomplemented in the MAJ operator, so it assigned to logic 1.

Variable e appears more often uncomplemented in the MAJ operator, so it assigned to logic 1.

The final assignment is then $(0, 1, 0, 1, 1)$ which evaluates $M(0, 1, 1, 1, 1, 1, 1, 1, 1) = 1$.

We have seen that the satisfiability of a single majority can be exactly decided in linear time, with respect to the size of the operator. The proposed greedy strategy is appropriate for such task. We show now how this procedure can be extended to handle two-level majority satisfiability.

5.5.2 Decide Strategy for MNF-SAT

For two-level MNF, a single *decide* may not be enough to determine SAT and it has to be iterated with *deduce* and *resolve* methods [8]. We propose here a *decide* strategy with linear time complexity with respect to the input MNF size. The rationale driving such process is to set each input variable to the logic value, 0 or 1, that maximizes the number of logic 1 in input to the final majority operator.[2] A corresponding automated procedure is depicted by Algorithm 8 and explained as follows. A specific variable x_j is first passed to the procedure, together with the MNF structure information. Then, a metric is computed to decide the assignment of such variable to logic 0 or 1. The main difference with respect to the one-level majority is indeed the figure of merit used to drive the variable assignment. The description of a proper metric is as follows. Say n the number (odd) of inputs of the final majority in an MNF. Thus, there are n majorities in the MNF. Say m_i the number (odd) of inputs of the ith majority operator, with $i \in \{1, 2, \ldots, n\}$. Say $n_p(x_j, i)$ the number of occurrence of variable x_j uncomplemented, in the ith majority operator. Similarly, say $n_c(x_j, i)$ the number of occurrence of variable x_j complemented, in the ith majority operator. Using these informations, two cost metrics $C_p(x_j)$ and $C_n(x_j)$ are created. Such cost metrics range from 0 to 1 and indicate how much a positive (C_p) or negative (C_n) polarity assignment of a variable contribute to set the MNF to logic 1. They are computed as

---

Algorithm 8 MNF-SAT *Decide* for a single variable

INPUT: Inputs: variable x_j, MNF structure
OUTPUT: Assignment for x_j most probably to SAT

   compute n_p, n_c;
   compute $C_p(x_j), C_n(x_j)$;
   if $C_p(x_j) < C_n(x_j)$ **then**
      $x_j = 0$;
   else
      $x_j = 1$;
   end if

---

[2]The final majority operator in an MNF is the one in the top layer of the two-level representation form, thus computing the output MNF function.

$C_p(x_j) = (\sum_{i=1}^{n} n_p(x_j, i)/m_i)/n$ and
$C_n(x_j) = (\sum_{i=1}^{n} n_c(x_j, i)/m_i)/n$.

According to this rationale, variable x_j is set to logic 1 if its positive polarity "convenience metric" $C_p(x_j)$ is greater than its negative polarity "convenience metric" $C_n(x_j)$. Otherwise, variable x_j is set to logic 0. Finally, a valid assignment for variable x_j is obtained. If iterated over all the variables, Algorithm 8 determines a global assignment to evaluate the MNF. Such procedure can be used as core *decide* task in a traditional *Decide - Deduce - Resolve* SAT solving approach [8]. Note that also the *deduce* and *resolve* methods must be adapted to the MNF nature. Although, new and *ad hoc deduce* and *resolve* techniques are desirable, their study is out of the scope of the current chapter. A simple example for the *decide* task, iterated over all the variables, is provided hereafter.

Example 5.12 We want to determine the satisfiability for the MNF formula $M(M(a, b, c', d, 1), M(a, b', c', d, e'), M(a', b, 0))$. Variables are considered in the order (a, b, c, d, e) and their cost metrics are computed.

For variable a, $C_p(a) = 1/5 + 1/5 + 0/3 = 0.4 > C_n(a) = 0/5 + 0/5 + 1/3 = 0.33$, thus it is assigned to logic 1.

For variable b, $C_p(b) = 1/5 + 0/5 + 1/3 = 0.53 > C_n(b) = 0/5 + 1/5 + 0/3 = 0.2$, thus it is assigned to logic 1.

For variable c, $C_p(c) = 0/5 + 0/5 + 0/3 = 0 < C_n(c) = 1/5 + 1/5 + 0/3 = 0.4$, thus it is assigned to logic 0.

For variable d, $C_p(d) = 1/5 + 1/5 + 0/3 = 0.4 > C_n(d) = 0/5 + 0/5 + 0/3 = 0$, thus it is assigned to logic 1.

For variable e, $C_p(e) = 0/5 + 0/5 + 0/3 = 0 < C_n(e) = 0/5 + 1/5 + 0/3 = 0.2$, thus it is assigned to logic 0.

The obtained assignment is then $(1, 1, 0, 1, 0)$ which evaluates $M(M(1, 1, 1, 1, 1), M(1, 0, 1, 1, 1), M(0, 1, 0)) = M(1, 1, 0) = 1$. The initial MNF formula is declared satisfiable.

Even though a single iteration may not be enough to determine the satisfiability of an MNF, the proposed *linear time decide* procedure can be used as core engine in a traditional SAT flow.

In the following section, we discuss the results obtained so far and highlight future research directions.

5.6 Discussion and Future Work

Two-level logic representation and satisfiability are two linked problems that have been widely studied in the past years. Nevertheless, the research in this field is still active. New approaches are continuously discovered and embedded in tools [12], to push further the horizons of logic applications. The proposed MNF has the potential to enhance two-level logic representation and related SAT problems.

We demonstrated that any CNF or DNF can be translated in linear time into an MNF. However, in its unrestricted form, MNF leads to SAT problems as difficult as with CNF. Restricted versions of MNF exist, whose satisfiability can be decided in polynomial time. Advanced logic manipulation techniques capable to transform a general MNF into a restricted MNF can significantly simplify the MNF-SAT problem. Also, direct MNF construction from general logic circuits is of interest.

Regarding the MNF representation properties, it is still unclear whether a canonical form exists for MNF, as it does for CNF (product of maxterms) and DNF (sum of minterms). The discovery of a canonical MNF can reveal new promising features of majority logic.

In the context of MNF-SAT algorithms, a detailed study for MNF oriented *deduce* and *resolve* techniques is required. In this way, a complete MNF-SAT solver can be developed and its efficiency tested.

In summary, our next efforts are focused on (i) logic manipulation techniques for MNF, (ii) canonical MNF representation, (iii) MNF-oriented *deduce* and *resolve* techniques and (iv) development of an MNF-SAT tool.

5.7 Summary

We presented, in this chapter, an alternative two-level logic representation form based solely on majority and complementation operators. We called it *Majority Normal Form* (MNF). MNF is *universal* and potentially more compact than its CNF and DNF counterparts. Indeed, MNF includes both CNF and DNF representations. We studied the problem of MNF-*SATisfiability* (MNF-SAT) and we proved that it belongs to the NP-complete complexity class, as its CNF-SAT counterpart. However, we showed practical restrictions on MNF formula whose satisfiability can be decided in polynomial time. We have finally proposed a simple core procedure to solve MNF-SAT, based on the intrinsic functionality of two-level majority logic. The theory and techniques developed in this chapter set the basis for future research on MNF-SAT solving.

References

1. L. Amaru, P.-E. Gaillardon, G. De Micheli, BDS-MAJ: A BDD-based logic synthesis tool exploiting majority logic decomposition, in *Proceedings of the DAC* (2013)
2. L. Amaru, P.-E. Gaillardon, G. De Micheli, Majority inverter graphs, in *Proceedings of the DAC* (2014)
3. G. Yang, W.N.N. Hung, X. Song, M. Perkowski, Majority-based reversible logic gates. Theor. Comput. Sci. **334**, 259–274 (2005)
4. H. Pospesel, *Introduction to Logic: Propositional Logic* (Pearson, New York, 1999)
5. T. Sasao, *Switching Theory for Logic Synthesis* (Springer, Heidelberg, 1999)
6. S. Cook, The complexity of theorem-proving procedures, in *Proceedings of ACM Symposium on Theory of Computing* (1971)

7. A. Biere, M. Heule, H. van Maaren, T. Walsh, *Handbook of Satisfiability* (IOS Press, Amsterdam, 2009)
8. K.J. Chen et al., InP-based high-performance logic elements using resonant-tunneling devices. IEEE Electr. Dev. Lett. **17**(3), 127–129 (1996)
9. M. Krause, P. Pudlak, On the computational power of depth-2 circuits with threshold and modulo gates. Theor. Comput. Sci. **174**, 137–156 (1997)
10. A.A. Sherstov, Separating AC 0 from depth-2 majority circuits, *Proceedings of the STOC* (2007)
11. N. Een, N. Sorensson, MiniSat - A SAT Solver with Conflict-Clause Minimization, SAT (2005)
12. N. Een, A. Mishchenko, N. Sorensson, Applying Logic Synthesis for Speeding Up SAT, SAT (2007)
13. R.E. Bryant, Graph-based algorithms for Boolean function manipulation. IEEE Trans. Comput. **C–35**, 677–691 (1986)
14. M.R. Garey, D.S. Johnson, *Computers and Intractability: A Guide to the Theory of NP-Completeness* (W. H. Freeman, San Francisco, 1979)
15. H.R. Lewis, Satisfiability problems for propositional calculi. Math. Syst. Theory **13**, 45–53 (1979)
16. R. Impagliazzo, A Satisfiability Algorithm for Sparse Depth Two Threshold Circuits, Arxiv (2013)

Chapter 6
Improvements to the Equivalence Checking of Reversible Circuits

Reversible circuits implement invertible logic functions. They are of great interest to cryptography, coding theory, interconnect design, computer graphics, quantum computing, and many other fields. As for conventional circuits, checking the combinational equivalence of two reversible circuits is an important but difficult (coNP-complete) problem. In this chapter, we present a new approach for solving this problem significantly faster than the state-of-the-art. For this purpose, we exploit inherent characteristics of reversible computation, namely bi-directional (invertible) execution and the XOR-richness of reversible circuits. Bi-directional execution allows us to create an *identity miter* out of two reversible circuits to be verified, which naturally encodes the equivalence checking problem in the reversible domain. Then, the abundant presence of XOR operations in the identity miter enables an efficient problem mapping into XOR-CNF satisfiability. The resulting XOR-CNF formulas are eventually more compact than pure CNF formulas and potentially easier to solve. As previously anticipated, experimental results show that our equivalence checking methodology is more than one order of magnitude faster, on average, than the state-of-the-art solution based on established CNF-formulation and standard SAT solvers.

6.1 Introduction

Reversible computing is a *non-conventional* computing style where all logic processing is conducted through bijective, i.e., invertible, Boolean functions. Reversible circuits implement invertible Boolean functions at the logic level and are represented as cascades of reversible gates. In conventional technologies, reversible circuits find application in cryptography [1], coding theory [2], interconnect design [3], computer graphics [4] and many other fields where the logic invertibility is a key asset. In emerging technologies, such as quantum computing [5], reversible circuits are one of the primitive computational building blocks.

© Springer International Publishing Switzerland 2017
L.G. Amaru, *New Data Structures and Algorithms for Logic Synthesis and Verification*, DOI 10.1007/978-3-319-43174-1_6

Whether they are finally realized in conventional or emerging technologies, the design of reversible circuits faces two major conceptual challenges: synthesis and verification [6]. Synthesis maps a target Boolean function into the reversible logic domain while minimizing the number of additional information bits and primitive gates [7, 8]. Verification checks if the final reversible circuit conforms to the original specification [9].

In this chapter, we focus on reversible circuit verification and, in particular, on combinational equivalence checking. The problem of combinational equivalence checking consists of determining whether two given reversible circuits are functionally equivalent or not. As for conventional circuits, this is a difficult (coNP-complete) problem [10]. We present a new approach for solving this problem significantly faster than the state-of-the-art verification approaches [9].

For this purpose, our methodology exploits, for the first time, inherent characteristics of reversible computation, i.e., its invertible execution and the XOR-richness of reversible circuits. This stands in contrast to previously proposed solutions such as introduced in [9] which only adapted established verification schemes for conventional circuits but ignored the potential of the reversible computing paradigm. Our proposed methodology consists of the following steps. First, we create an identity miter by cascading one circuit with the inverse of the other. If the two reversible circuits are functionally equivalent, then the resulting cascade realizes the identity function. Next, we encode the problem of checking whether the resulting circuit indeed realizes the identity into a mixed XOR-CNF satisfiability problem. The possibility to express natively XOR operations, frequently appearing in reversible circuits, reduces significantly the number of variables and clauses as compared to a pure CNF formulation. Finally, we solve the XOR-CNF satisfiability problem using CryptoMiniSat [11], a MiniSat-based solver handling XORs through Gaussian elimination [12]. Experimental results show that, on average, the proposed methodology is more than one order of magnitude faster than the state-of-the-art reversible circuit checker based on the established CNF-formulation and MiniSat solver [9]. Besides that, the proposed approach also provides potential for improving combinational equivalence checking of conventional circuits.

The remainder of this chapter is organized as follows. Section 6.2 provides the background on reversible circuits and on Boolean satisfiability. Section 6.3 presents the proposed methodology for equivalence checking of reversible circuits. Section 6.4 describes the setup applied for our experimental evaluation and summarizes the obtained results. Section 6.5 discusses the future research directions—in particular for combinational equivalence checking of conventional circuits. Section 6.6 concludes the chapter.

6.2   Background

In this section, we briefly review the background on reversible circuits and on Boolean satisfiability.

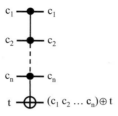

Fig. 6.1 A Toffoli gate

6.2.1 Reversible Circuits

A logic function $f : \mathbb{B}^{n_i} \to \mathbb{B}^{n_o}$ is *reversible* if and only if it represents a bijection. This implies that:

- the number of inputs is equal to its number of outputs (i.e., $n_i = n_o$) and
- it maps each input pattern to a unique output pattern.

A reversible function can be realized by a circuit $G = g_1 g_2 \cdots g_d$ comprised of a cascade of reversible gates g_i, where d is the number of gates. Multiple forks and feedback are not directly allowed [5]. Several different reversible gates have been introduced including the Toffoli gate [13], the Fredkin gate [14], and the Peres gate [15]. In accordance to the common approach in reversible circuit design (see e.g., [7, 8]), we focus on Toffoli gates in the following. Toffoli gates are universal, i.e., all reversible functions can be realized by means of this gate type alone [13].

A *Toffoli gate* has a *target line* t and *control lines* $\{c_1, c_2, \ldots, c_n\}$.[1] Its behavior is the following: If all control lines are set to the logic value 1, i.e., $c_1 \cdot c_2 \cdot \cdots \cdot c_n = 1$, the target line t is inverted, i.e., t'. Otherwise, the target line t is passed through unchanged. Hence, the Boolean function of the target line can be expressed as $(c_1 \cdot c_2 \cdot \cdots \cdot c_n) \oplus t$. All remaining signals (including the signals of the control lines) are always passed through unchanged. Figure 6.1 depicts a Toffoli gate with its respective output functions. We follow the established drawing convention of using the symbol $\oplus$ to denote the target line and solid black circles to indicate control connections for the gate.

A Toffoli gate with no control lines always inverts the target line and is a *NOT gate*. A Toffoli gate with a single control line is called a *controlled-NOT gate* (also known as the CNOT gate) and is functionally equivalent to a XOR gate. The case of two control lines is the original gate defined by Toffoli [13].

Example 6.1 Figure 6.2 shows a reversible circuit composed of $m = 3$ circuit lines and $d = 6$ Toffoli gates. This circuit maps each input pattern into a unique output pattern. For example, it maps the input pattern 111 to the output pattern 100. Inherently,

---

[1]Toffoli gates have bee briefly introduced in Chap. 2. Here, their functionality is presented for the sake of clarity.

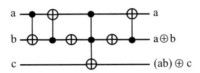

Fig. 6.2 A reversible circuit composed of Toffoli gates

every computation can be performed in both directions (i.e., computations towards the outputs *and* towards the inputs can be performed).

6.2.2   Boolean Satisfiability

The Boolean *Satisfiability* (SAT) problem has been defined and discussed in Chaps. 4 and 5 of this book, respectively. For the sake of clarity, we report here an example of *Conjunctive Normal Form* (CNF)-SAT as:

$$(a + b')(a + c')(a' + b + c)$$

which is satisfiable by $(a = 1, b = 1, c = 1)$.

Even though SAT for generic CNFs is a difficult (NP-complete) problem, modern SAT solvers can handle fairly large problems in reasonable time [16]. The core technique behind most SAT solvers is the DPLL (Davis-Putnam-Logemann-Loveland) procedure, introduced several decades ago [17]. It performs a backtrack search in the space of partial truth assignments. Through the years, the main improvements to DPLL have been smart branch selection heuristics, a fast implication scheme, and extensions such as clause learning, randomized restarts, as well as well-crafted data structures such as lazy implementations and watched literals for fast unit propagation [16].

Recently, researchers considered SAT to solve other important problems in computer science, for example, cryptographical applications [19]. Here, SAT solvers are often faced with a large amount of XOR constraints. These XORs are typically difficult to handle using pure CNF and standard SAT solvers. However, the presence of these XOR constraints can be exploited within a DPLL solving framework by using on-the-fly Gaussian elimination [12]. Some SAT solvers have been proposed which exploit this potential and, hence, work on mixed XOR-CNF formulas rather than pure CNF formulas. For example, a mixed XOR-CNF is

$$(a \oplus b')(a \oplus c)(a' + b + c)$$

which is satisfiable by $(a = 1, b = 1, c = 0)$.

CryptoMiniSat [11] is one of the most popular solvers for XOR-CNF formulas based on MiniSat [24] and Gaussian elimination to handle XOR constraints [12].

6.3 Mapping Combinational Equivalence Checking for Reversible Circuits to XOR-CNF SAT

In this section, we present the proposed approach for checking the combinational equivalence between two reversible circuits. Without loss of generality, we consider reversible circuits composed only of Toffoli, CNOT, and NOT gates. Since Toffoli gates are universal, any other primitive reversible gate can be decomposed into a combination of those.

In the remainder of this section, we first describe how to create an *identity miter* out of two reversible circuits under test. Then, we propose an efficient encoding of the identity check problem into XOR-CNF satisfiability.

6.3.1 Creating an Identity Miter

In the considered scenario, two reversible circuits need to be checked for combinational equivalence. As an example, consider the circuits depicted in Fig. 6.3.

Following established verification schemes, both circuits are fed by the same input signals. Differences at the outputs are observed by applying XOR operations. This eventually lead to a new circuit specifically used for equivalence checking which is commonly called *miter circuit* [20]. If at least one output of the miter can evaluate

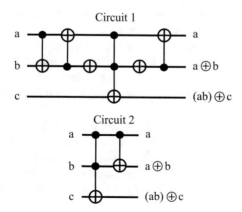

Fig. 6.3 Two functionally equivalent reversible circuits

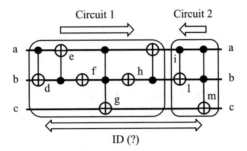

Fig. 6.4 The resulting identity miter

to the logic value 1, for some input pattern, then the two circuits are functionally different. Otherwise, the two circuits are functionally equivalent.

The very same approach can be used to verify the combinational equivalence of two reversible circuits (and, in fact, has been done before in [9]). However, just an adaptation of this conventional scheme entirely ignores the potential that comes by following the reversible computing paradigm. In fact, properties of reversible circuits can be exploited to create a different type of miter. More precisely, a reversible circuit realizes a function f when considered from the inputs to the outputs. But thanks to the reversibility, it also realizes the inverse function f^{-1} when considered from the outputs to the inputs.[2] Therefore, by cascading one reversible circuit with the inverse (I/O flip) of a functional equivalent one always yields to a circuit realizing the identity function over all signal lines. This concept is illustrated in Fig. 6.4 which shows the resulting identity miter comprised from the example circuits of Fig. 6.3.

We call such composite circuit an *identity miter*. If at least one output of the identity miter does not represent the identity function, i.e., if $f(x) \neq x$, then the two reversible circuits are functionally different. Otherwise, the two circuits are functionally equivalent.

Note that the idea of creating an identity miter out of two reversible circuits is not new per se. Indeed, it has been already studied in [18]. However, in that work, the use of an identity miter did not lead to substantial improvements for equivalence checking of reversible circuits. This is because researchers used canonical data structures, decision diagrams and alike, to perform the identity checking task. The scaling limitations of canonical data structures severely confined the potential efficiency of using an identity miter.

Instead, in this work, we propose an innovative SAT formulation to describe the identity miter checking problem. SAT can handle much larger problems than canonical data structures before hitting serious scaling limitations. Moreover, we develop an ad-hoc mixed XOR-CNF formulation to natively handle the identity miter checking problem and significantly expedite its solving as compared to a pure CNF formulation.

---

[2]This holds since self-inverse reversible gates such as Toffoli gates, CNOT gates, NOT gates, etc. are considered here.

6.3.2 XOR-CNF Formulation

To test the equivalence of two reversible circuits, we need to check whether their *identity miter* actually represents an identity function or not. If such an assignment can be determined, then the identity miter does not actually represents the identity function and the two reversible circuits under test are not functionally equivalent (in this case, the determined assignment works as counterexample). Otherwise, the identity miter represents the identity function and the two reversible circuits are functionally equivalent.

Besides that, the XOR-richness of the considered circuits can be exploited. In fact, most of the reversible circuits are inherently composed of XOR operations only—caused by the applied Toffoli gate library as introduced in Sect. 6.2.1. This allows for a formulation in terms of a mixed XOR-CNF satisfiability problem which, as reviewed in Sect. 6.2.2, can be handled much better using dedicated solvers rather than the conventionally applied CNF satisfiability.

The resulting formulation is defined as follows: First, corresponding SAT variables are introduced. More precisely, for each primary input of the identity miter as well as for each reversible gate, a new free variable is introduced.

Example 6.2 Consider again the identity miter as shown in Fig. 6.4. For the primary inputs, the variables a, b, c are introduced. The variables $d, e, \ldots, m$ represent reversible gates outputs.

Afterwards, two types of constraints are introduced: The first type covers the functionality of the circuit, i.e., symbolically restricts the set of possible assignments to those which are valid with respect to the given gate functions and connections. The second type covers the objective, i.e., symbolically restricts the set of possible assignments to those which show, for at least one circuit line, the non-identity of the input x and the output $f(x)$ (in other words, assignments which violate $x = f(x)$).

Considering the functional constraints, there are as many functional constraints as Toffoli, CNOT, and NOT gates in the circuit. Each of them introduces its particular set of functional constraints which restrict the output value (denoted by o in the following) of the respective target lines. More precisely,

- a NOT gate with a target line t is represented by $(o = t')$,
- a CNOT gate with target line t and control line c is represented by $(o = c \oplus t)$, and
- a Toffoli gate with target line t and control lines $\{c_1, c_2, \ldots, c_n\}$ is represented by $(o = p \oplus t)$ and $(p = c_1 \cdot c_2 \cdot \cdots \cdot c_n)$.

All these constraints must simultaneously hold in order to properly represent the circuit functionality.

Example 6.3 Consider again the identity miter shown in Fig. 6.4. For this circuit, the following functional constraints are created:

$$\text{Functionality} \begin{cases} d = b \oplus a \\ e = a \oplus d \\ f = d' \\ g = c \oplus p_1 \\ p_1 = f \cdot e \\ h = f' \\ i = e \oplus h \\ l = h \oplus i \\ m = g \oplus p_2 \\ p_2 = i \cdot l \end{cases} \tag{6.1}$$

As an example, consider the variable g which symbolically represents the output value of the fourth gate from Fig. 6.4. The functionality of this gate is represented by $g = c \oplus p_1$. The variable p_1 represents thereby the controlling part of this Toffoli gate and is accordingly represented as $p_1 = f \cdot e$. The remaining constraints in Eq. 6.1 are derived analogously.

Considering the objective constraints, there are as many objective constraints as lines in the reversible circuit. Here, the functional constraints as described above are utilized. For a generic $line_i$ ($i \leq m$), the primary outputs in the identity circuit are respectively defined by the cascade of gates $g_1 g_2 \cdots g_d$. The functional constraints represent these gates by means of a cascade of XOR operations so that $line_i$ is eventually defined as $line_i = h_1 \oplus h_2 \cdots \oplus h_d$ where each h_j ($j \leq d$) is either

- the product $p = c_1 \cdot c_2 \cdot \cdots \cdot c_n$ of the control connections of gate g_j (in case the corresponding gate g_j is a Toffoli gate),
- the control signal c (in case the corresponding gate g_j is a CNOT gate), or
- the logic value 1 (in case the corresponding gate g_j is a NOT gate).

Because of this cascade of XOR operations, the objective constraints only have to ensure that, for at least one $line_i$, its corresponding output assumes the logic value 1, i.e., behaves as an inverter rather than a buffer. This can be formulated as $\exists i \in \{1, 2, \ldots, m\} : line_i = 1$, where m is the number of lines.

Example 6.4 Consider again the identity circuit considered above. For this circuit, the following objective constraints are added:

$$\text{Non-Identity} \begin{cases} \exists i \in \{1, 2, 3\} : line_i = 1 \\ line_1 = d \oplus h \\ line_2 = a \oplus 1 \oplus 1 \oplus i \\ line_3 = p_1 \oplus p_2 \end{cases} \tag{6.2}$$

As an example, consider the bottom (third) line of the reversible circuit from Fig. 6.4. We have that $line_3 = p_1 \oplus p_2$. The values of p_1 and p_2 are derived from the functional constraints, in particular from control lines of the respective Toffoli gates. The objective constraint asks for at least one of the three lines to evaluate to the logic value 1, thus to invert the corresponding input bit (so not being an identity).

As one can visually notice, the set of constraints in Eqs. 6.1 and 6.2 are not yet in XOR-CNF form. Hence, some further transformations are needed. For this purpose, we exploit the fact that, in the Boolean domain, $(a = b)$ can equally be represented as $(a \oplus b' = 1)$. This allows us to transform most of the equalities directly into XOR clauses. In contrast, special treatment is required for the AND constraints caused by the representations of the control lines, i.e., for p. For these ones, it is more efficient to rely on the established Tseitin transformation [21]. Tseitin transformation sets a particular gate Boolean expression equal to constant 1 and transforms it into a conjunction of disjunctions. For this reason, Tseitin transformation encodes an AND function over k inputs into $k + 1$ OR clauses. Finally, the constraint $\exists i \in \{1, 2, 3\}$: $line_i = 1$ is naturally mapped into a standard OR clause.

Example 6.5 Following the example from above, all constraints from Eqs. 6.1 and 6.2 are eventually transformed into the following single set of XOR-CNF clauses:

$$
\text{XOR-CNF}
\begin{cases}
d' \oplus b \oplus a \\
e' \oplus a \oplus d \\
f' \oplus d' \\
g' \oplus c \oplus p_1 \\
p_1 + f' + e' \\
p_1' + f \\
p_1' + e \\
h \oplus f \\
i' \oplus e \oplus h \\
l' \oplus h \oplus i \\
m' \oplus g \oplus p_2 \\
p_2 + i' + l' \\
p_2' + i \\
p_2' + l \\
line_1' \oplus d \oplus h \\
line_2' \oplus a \oplus 1 \oplus 1 \oplus i \\
line_3' \oplus p_1 \oplus p_2 \\
line_1 + line_2 + line_3
\end{cases}
\tag{6.3}
$$

The resulting XOR-CNF problem is unsatisfiable as the considered identity miter shown in Fig. 6.4 indeed represents the identity. That means that the two original reversible circuits to be verified (shown in Fig. 6.3) are combinationally equivalent. This can be proved manually or, more efficiently, using a XOR-CNF satisfiability solver.

Note that the XOR-CNF formulation in Eq. 6.3 is composed of 18 clauses and 16 variables. In contrast, the established formulation based on pure CNF requires 82 clauses and 34 variables [9]. This reduction alone is likely to lead to a solving speed-up. Moreover, the presence of more than 60 % XOR clauses opens even more speed-up opportunities. Mixed XOR-CNF solvers take advantage of XOR clauses through fast Gaussian elimination. Results showed in the next section confirm the predicted improvement.

6.4 Experimental Results

In order to evaluate the performance of the proposed approach, we implemented the techniques described above and compared them against the state-of-the-art solution presented in [9]. In this section, we summarize the respectively obtained results. Details on the applied methodology as well as the experimental setup are provided.

6.4.1 *Methodology and Setup*

The proposed equivalence checking scheme has been implemented as a tool chain which is sketched by Fig. 6.5. Two reversible circuits (provided in the *.real-format [23]) are taken and re-arranged into an identity circuit as well as mapped into an equivalent XOR-CNF formulation. For this purpose, the concepts described in Sect. 6.3 have been implemented in terms of a C-program. Afterwards, the resulting formulation is passed to CryptoMiniSAT 2.0—an XOR-CNF solver [11]. In case the solver proved the unsatisfiability of the instance, equivalence (EQ) has been proven; otherwise, it has been shown that the considered circuits are not equivalent (NEQ).

For comparison, we additionally considered the SAT-based reversible circuit checker presented in [9]. From a high-level perspective, this tool first creates an XOR-miter of the given reversible circuits. Then, it encodes the XOR-miter into a pure CNF formula which is eventually solved using MiniSAT [24]. Even though

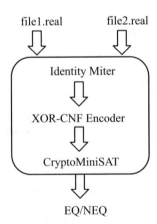

Fig. 6.5 The proposed equivalence checking flow

this flow has been explicitly tuned for verification of reversible circuits in [9], it still employs the state-of-the-art schemes as applied for verification of conventional circuits. To enable a fair runtime comparison, we downloaded, compiled, and run the reference tool from [9] for our evaluations.

As benchmarks, we considered reversible circuits (provided in the *.real-format) from the RevLib benchmark library [23]. We neglected small reversible circuits for which the verification runtime was less than a second. We focused on complex reversible circuits (with >2 k gates) for which the verification task required more computational effort. In particular, we give results for two classes of benchmarks: circuits realizing *Unstructured Reversible Functions* (URF) as well as circuits realizing arithmetic components of a RISC CPU. These classes are the largest and toughest benchmarks available at RevLib [23] and, hence, are appropriate to challenge the proposed verification scheme.

Whenever required, all gates in these circuits have been locally transformed into universal Toffoli gates. In order to consider both cases of equivalence as well as non-equivalence three versions of each circuit have been considered, namely (i) the original version, (ii) an optimized version, and (iii) an erroneous version.

All experiments have been conducted on a Dual Xeon 6 cores X5650 machine with 24 GB RAM running under RHEL 5.8–64 bits OS.

6.4.2 Results

Table 6.1 summarizes the experimental results. Considering the URF-benchmarks, equivalence checking can be conducted approximatively 9 times faster compared to the reference verification scheme. If the CPU-benchmarks are considered, even better improvements can be observed; namely speed-ups of a factor of approximatively 12. Here, particular the benchmark alu_5 is of interest. Applying the reference scheme proposed in [9], no result was obtained within 24 h (its contribution to the total runtime nevertheless has been considered as 24 h, i.e., 86400 in favor to the reference flow). In contrast, the proposed approach was able to check the equivalence in less than two hours. Over all benchmarks, an improvement of more than one order of magnitude (more precisely, a factor of 12.44) is observable.

We see the two reasons for this significant improvement: On the one hand, the number of variables and clauses are considerably smaller in the proposed XOR-CNF formulation compared to the pure CNF formulation (a reduction by the factor of 1.29 and 2.42, respectively). On the other hand, the richness of XOR-clauses in our formulation helps the solving engine in simplifying the formula early in the process (e.g., through Gaussian elimination). Further investigation is needed to numerically separate the contributions for each speedup source.

Besides that, non-equivalent cases have been solved quite faster than equivalent cases for both, the proposed scheme as well as the reference scheme. This is expected as SAT solvers are known to be very fast in detecting satisfying assignments rather than proving unsatisfiability.

Table 6.1 Experimental results (all run-times in CPU seconds)

Circuit 1 (lines/gates)	Circuit 2 (lines/gates)	State-of-the-art [9]			Proposed solution			
		Vars/Clauses	Answer	Runtime	Vars/Clauses	XOR %	Answer	Runtime
Unstructured Reversible Functions (from RevLib)								
urf3_1 (10/26 k)	urf3_2 (10/26 k)	133609/527485	EQ	98.85	104212/210085	32	EQ	14.20
urf3_1 (10/26 k)	urf3_bug (10/26 k)	133433/526926	NEQ	5.91	104212/210085	32	NEQ	1.69
urf1_1 (9/11 k)	urf1_2 (9/6 k)	58122/229437	EQ	17.89	35847/61885	60	EQ	2.54
urf1_1 (9/11 k)	urf1_bug (9/6 k)	58124/229390	NEQ	2.77	45438/91655	31	NEQ	0.52
urf5_1 (10/10 k)	urf5_2 (10/10 k)	51746/20401	EQ	15.85	40350/81455	31	EQ	3.75
urf5_1 (10/10 k)	urf5_bug (10/9 k)	51810/204249	NEQ	1.54	40312/81377	31	NEQ	0.42
urf6_1 (15/10 k)	urf6_2 (15/10 k)	54888/216888	EQ	5694.22	42565/85526	33	EQ	570.39
urf6_1 (15/10 k)	urf6_bug (15/9 k)	54682/216370	NEQ	2.64	42524/85445	33	NEQ	0.49
urf4_1 (11/32 k)	urf4_2 (11/31 k)	162247/636237	EQ	883.27	127254/255271	55	EQ	92.37
urf4_1 (11/32 k)	urf4_bug (11/31 k)	162349/636563	NEQ	6.04	127245/255252	55	NEQ	2.03
Total URF		921010/3443964	–	6728.98	709959/1418036	39	–	688.40
Components of the RISC CPU (from RevLib)								
alu1_1 (756/3 k)	alu1_2 (756/10 k)	82617/281684	EQ	5649.74	23128/99803	28	EQ	670.96
alu1_1 (756/3 k)	alu1_bug (756/2 k)	66182/216644	NEQ	67.84	10625/65921	21	NEQ	6.96
alu2_1 (6204/3 k)	alu2_2 (6204/3 k)	5568/20216	EQ	304.65	21254/22521	79	EQ	186.44
alu2_1 (6204/3 k)	alu2_bug (6204/3 k)	5657/20610	NEQ	369.49	21250/22517	80	NEQ	76.21
alu3_1 (255/10 k)	alu3_2 (255/11 k)	227505/752851	EQ	12751.02	35406/253957	12	EQ	728.98
alu3_1 (255/10 k)	alu3_bug (155/8 k)	209887/691117	NEQ	56.91	30424/232584	12	NEQ	9.90
alu4_1 (757/4 k)	alu4_2 (757/7 k)	28671/111480	EQ	8899.70	20941/40794	42	EQ	320.87
alu4_1 (757/4 k)	alu4_bug (757/4 k)	22140/85537	NEQ	825.71	16496/30987	55	NEQ	169.00
alu5_1 (256/9 k)	alu5_2 (256/10 k)	47290/185110	?	>1 day	33150/65249	45	EQ	6948.86
alu5_1 (256/9 k)	alu5_bug (256/9 k)	43966/171863	NEQ	51.56	30894/60329	51	NEQ	10.36
Total RISC CPU		739483/2537112	–	115376.62	243568/894662	42	–	9128.54
Grand total		1660493/5981058	–	122105.60	953527/2312698	41	–	9816.94
Improvement compared to [9]	1/1	–		1	**1.74/2.58×**	–	–	**12.44×**

6.5 Discussion

The proposed solution provides an alternative verification scheme for reversible logic which leads to significant improvements with respect to the state-of-the-art. Beyond that, it also opens promising new paths for improving verification of conventional designs. This section briefly discusses new research opportunities in this direction.

6.5.1 Application to the Verification of Conventional Circuits

The significant speed-up obtained in this work is enabled by intrinsic properties of reversible circuits such as bi-directional execution and XOR-richness. Conventional circuits usually do not inherit these particular properties. Hence, at a first glance, the proposed verification scheme may seem applicable only to reversible computation paradigms.

However, conventional logic can also be represented in terms of reversible logic by using extra I/Os and extra gates. Previous studies explored this direction in order to (ideally) map any combinational design into reversible circuits [25]. This motivates us to consider a *new verification flow*. The core idea is to convert convential circuits into reversible ones and perform the verification tasks in the reversible domain. In this way, the efficiency of the reversible equivalence checking flow proposed in this work can be further exploited. The conventional-to-reversible mapping may also be inefficient, from an optimization standpoint, but the benefits demonstrated so far are large enough to absorbe such inefficiency and possibly leave room for a relevant improvement.

The main issue here is defining a robust and trustable conventional-to-reversible mapping technique. In this context, existing conventional-to-reversible mapping techniques [26] do not natively fit the requirements as they are intrinsically developed for logic optimization purposes. Our future research efforts are focused on the development of such reversible conversion method starting from arbitrary combinational logic circuits.

Provided that, traditional verification tasks will take full advantage of the reversible computing paradigm opening new exciting research directions.

6.5.2 Easy Exploitation of Parallelism

The proposed equivalence checking method can be further improved by exploiting concurrent execution. To speed-up SAT solvers, researchers are studying parallel and concurrent execution (e.g., [27]). This is motivated by the fact that, nowadays, multi-cores are wide-spread and computing resources are inexpensive. However, to fully exploit the potential offered by parallelization, also the respective SAT problems must be formalized in a parallel fashion. This is usually not obvious for the established equivalence checking solutions proposed in the past.

In contrast, a parallel consideration is simple for the solution proposed in this work. Indeed, the formulation described in Sect. 6.3 can easily be split for each circuit line. By this, the overall equivalence checking problem is decomposed into m separate instances (with m being the number of circuit lines). These instances are smaller and can be solved independently from each other. As soon as one of the instances is found satisfiable, non-equivalence has been proven. Overall, this does not only allow for easier instances to be separately solved, but also enables the full exploitation of multiple-cores—something which is much harder to accomplish for almost all (conventional) verification schemes available thus far.

6.6 Summary

Reversible circuits are of great interest to various fields, including cryptography, coding theory, communication, computer graphics, quantum computing, and many others. Checking the combinational equivalence of two reversible circuits is an important but difficult (coNP-complete) problem. In this chapter, we presented a new approach for solving this problem significantly faster than the state-of-the-art. The proposed methodology explicitly exploited the inherent properties of reversible circuits, namely the bi-directional execution as well as their XOR-richness. This eventually enabled speed-ups of more than one order of magnitude on average. While this represents a substantial improvement for the verification of circuit descriptions aimed for reversible computation, it also offers promising potential to be exploited in the verification of conventional designs.

References

1. D. Kamalika, I. Sengupta, Applications of reversible logic in cryptography and coding theory, in *Proceedings of the 26th International Conference on VLSI Design* (2013)
2. K. Czarnecki et al., Bidirectional transformations: a cross-discipline perspective, in *Theory and Practice of Model Transformations*, ed. by R.F. Paige (Springer, Berlin, 2009), pp. 260–283
3. R. Wille, R. Drechsler, C. Oswald, A. Garcia-Ortiz, Automatic design of low-power encoders using reversible circuit synthesis, in *Design, Automation and Test in Europe (DATE)* (2012), pp. 1036–1041
4. S.L. Sunil, C.D. Yoo, T. Kalker, Reversible image watermarking based on integer-to-integer wavelet transform. IEEE Trans. Inf. Forensics Secur. **2**(3), 321–330 (2007)
5. M. Nielsen, I.L. Chuang, *Quantum Computation and Quantum Information* (Cambridge University Press, Cambridge, 2010)
6. R. Wille, R. Drechsler, *Towards a Design Flow for Reversible Logic* (Springer, Berlin, 2010)
7. R. Drechsler, R. Wille, From truth tables to programming languages: Progress in the design of reversible circuits, in *International Symposium on Multiple-Valued Logic* (2011), pp. 78–85
8. M. Saeedi, I.L. Markov, Synthesis and optimization of reversible circuits–a survey. ACM Comput. Surv. (CSUR) **45**(2), 21 (2013)

9. R. Wille et al., Equivalence checking of reversible circuits, in *39th IEEE International Sympo-sium on Multiple-Valued Logic* (2009)

10. S.P. Jordan, Strong equivalence of reversible circuits is coNP-complete. Quantum Inf. Comput. **14**(15–16), 1302–1307 (2014)

11. CryptoMiniSAT tool. http://www.msoos.org/cryptominisat2/

12. M. Soos, Enhanced Gaussian Elimination in DPLL-based SAT Solvers POS@SAT (2010)

13. T. Toffoli, Reversible computing, in *Automata, Languages and Programming*, ed. by W. de Bakker, J. van Leeuwen (Springer, Heidelberg, 1980), p. 632. (Technical Memo MIT/LCS/TM-151, MIT Lab. for Comput. Sci.)

14. E.F. Fredkin, T. Toffoli, Conservative logic. Int. J. Theor. Phys. **21**(3/4), 219–253 (1982)

15. A. Peres, Reversible logic and quantum computers. Phys. Rev. A **32**, 3266–3276 (1985)

16. A. Biere, M. Heule, H. van Maaren (eds.) *Handbook of Satisfiability*, vol. 185 (IOS Press, Amsterdam, 2009)

17. M. Davis, G. Logemann, D. Loveland, A machine program for theorem proving. Commun. ACM **5**(7), 394–397 (1962)

18. S. Yamashita, I.L. Markov. Fast equivalence-checking for quantum circuits, in *Proceedings of the 2010 IEEE/ACM International Symposium on Nanoscale Architectures* (IEEE Press, Piscataway, 2010)

19. M. Soos, K. Nohl, C. Castelluccia, *Extending SAT solvers to cryptographic problems, Theory and Applications of Satisfiability Testing-SAT 2009* (Springer, Berlin, 2009), pp. 244–257

20. D. Brand, Verification of large synthesized designs, in *Proceedings of the ICCAD* (1993), pp. 534–537

21. G.S. Tseitin, *On the Complexity of Derivation in Propositional Calculus, Automation of Rea-soning* (Springer, Berlin, 1983)

22. Reversible CEC flow and experiments of this work. http://lsi.epfl.ch/RCEC

23. R. Wille, D. Große, L. Teuber, G.W. Dueck, R. Drechsler, RevLib: an online resource for reversible functions and reversible circuits, in *International Symposium on Multiple-Valued Logic*, RevLib is (2008). http://www.revlib.org, pp. 220–225

24. MiniSat: open-source SAT solver. http://minisat.se

25. D.M. Miller, R. Wille, G.W. Dueck, Synthesizing reversible circuits for irreversible functions, in: *Euromicro Conference on Digital System Design (DSD)* (2009), pp. 749–756

26. R. Wille, O. Keszöcze, R. Drechsler, Determining the minimal number of lines for large reversible circuits, in *Design, Automation and Test in Europe (DATE)* (2011)

27. Y. Hamadi et al., ManySAT: a parallel SAT solver. J. Satisf. Boolean Model. Comput. **6**, 245–262 (2008)

Chapter 7
Conclusions

In this book, we investigated new data structures and algorithms for *Electronic Design Automation* (EDA) logic tools, in particular for logic synthesis and verification. Motivated by (i) the ever-increasing difficulty of keeping pace with design goals in modern CMOS technology and (ii) the rise of enhanced-functionality nanotechnologies, we studied novel logic connectives and Boolean algebra extending the capabilities of synthesis and verification techniques. The results presented in this book give an affirmative answer to the question *"Can EDA logic tools produce better results if based on new, different, logic primitives?"*.

7.1 Overview of Book Contributions

The overview proceeds following the order of the presentation.

- **We improved the efficiency of logic representation, manipulation and optimization tasks by taking advantage of majority and biconditional logic expressiveness**. Majority logic is a powerful generalization of standard AND/OR logic. Biconditional logic intrinsically realizes an equality check over Boolean variables. Majority and biconditional connectives together form the basis for arithmetic logic. We developed synthesis techniques exploiting majority and biconditional logic properties [1–3]. Our tools showed strong results as compared to state-of-the-art academic and commercial synthesis tools. Indeed, we produced the best (public) results for many circuits in combinational benchmark suites [4]. On top of the enhanced synthesis power, our methods are also the natural and native logic abstraction for circuit design in emerging nanotechnologies, where majority and biconditional logic are the primitive gates for physical implementation [5].

© Springer International Publishing Switzerland 2017
L.G. Amaru, *New Data Structures and Algorithms for Logic
Synthesis and Verification*, DOI 10.1007/978-3-319-43174-1_7

- **We accelerated formal methods by (i) studying core properties of logic circuits and (ii) developing new frameworks for logic reasoning engines**. Thanks to the majority logic representation theory, we discovered non-trivial dualities in the property checking problem for logic circuits [6]. Our findings enabled sensible speed-ups in solving circuit satisfiability. With the aim of exploiting further the expressive power of majority logic, we developed an alternative Boolean satisfiability framework based on majority functions [7]. We proved that the general problem is still intractable but we showed practical restrictions that instead can be solved efficiently. Finally, we focused on the important field of reversible logic and we proposed a new approach to solve the equivalence checking problem [8]. We defined a new type of reversible miter over which the equivalence check test is performed. Also, we represented the core checking problem in terms of biconditional logic. This enabled a much more compact formulation of the problem as compared to the state-of-the-art. Indeed, it translated into more than one order of magnitude speed up for the overall application, as compared to the state-of-the-art solution.

7.2 Open Problems

We give some directions for future research.

- **Theoretical study on the size of biconditional binary decision diagrams for notable functions**. Multiplier and hidden-weight bit functions are represented by exponential sized BDDs, no matter what variable order is employed. It would be interesting to study their size in BBDD representation and prove the gap, if any, with respect to other DD representations.
- **Majority-biconditional logic manipulation**. A single data structure merging biconditional and majority logic together can improve even further the efficiency of logic synthesis. Indeed, majority and biconditional share interesting properties, e.g., the propagation of biconditional operators into majority operators and viceversa. Moreover, majority and biconditional together are the natural basis for arithmetic logic. It would be interesting to study the properties of majority-biconditional logic manipulation, especially in light of its application to arithmetic function synthesis.
- **Exact majority logic synthesis**. In contrast to heuristic methods, exact synthesis methods determine a minimal circuit implementation in terms of either number of gates or number of levels. State-of-the-art exact synthesis methods, for AND/OR logic circuits, deal with functions up to 5 variables by means of smart enumeration techniques. It would be interesting to exploit Boolean properties of majority logic, e.g., orthogonal errors masking, to design an exact depth optimization method for MIGs which pushes further the exact synthesis complexity frontier.
- **MNF-SAT solver**. A practical MNF-SAT solver has yet to be developed together with *ad-hoc deduce* and *resolve* techniques for majority logic.

- **Reversible equivalence checking for conventional designs**. Conventional logic can also be represented in terms of reversible logic by using extra I/Os and extra gates. Previous studies explored this direction in order to (ideally) map any combinational design into reversible circuits. In this scenario, it would be interesting to apply the reversible equivalence checking paradigm to conventional designs post-converted into reversible circuits. If the conventional-to-reversible mapping is efficient enough, the speedup observed in Chap. 6 is likely to appear also for the verification of conventional designs.

7.3 Concluding Remarks

In this book, we approached fundamental EDA problems from a different, unconventional, perspective. Our synthesis and verification results demonstrate the key role of rethinking EDA solutions in overcoming technological limitations of present and future technologies.

In addition to new EDA studies, this book opens also other research paths. For example, MIG logic manipulation can speed up Big Data processing via better mappings of (high-performance) programming languages. Also, BBDDs can model natively modulo (encrypting) operations in Data Security applications making secure computation more efficient. We believe the material presented in this book will prove useful in these and many other fields of computer science.

References

1. L. Amaru, P.-E. Gaillardon, G. De Micheli, Majority Inverter Graphs, Proceedings, DAC (2014)
2. L. Amaru, P.-E. Gaillardon, G. De Micheli, *Boolean Optimization in Majority Inverter Graphs* (Proc, DAC, 2015)
3. L. Amarù, P.-E. Gaillardon, G. De Micheli, Biconditional binary decision diagrams: a novel canonical representation form. IEEE J. Emerg. Sel. Top. Circuits Syst. (JETCAS) 4(4), 487–500 (2014)
4. The EPFL Combinational Benchmark Suite. http://lsi.epfl.ch/benchmarks
5. L. Amarù, P.-E. Gaillardon, S. Mitra, G. De Micheli, New logic synthesis as nanotechnology enabler, in *Proceedings of the IEEE* (2015)
6. L. Amarù, P.-E. Gaillardon, A. Mishchenko, M. Ciesielski, G. De Micheli, Exploiting circuit duality to speed Up SAT, in *IEEE Computer Society Annual Symposium on VLSI (ISVLSI), Montpellier* (2015)
7. L. Amarù, P.-E. Gaillardon, G. De Micheli, Majority logic representation and satisfiability, in *23rd International Workshop on Logic and Synthesis (IWLS), San Francisco* (2014)
8. L. Amaru, P.-E. Gaillardon, R. Wille, G. De Micheli, in *Exploiting Inherent Characteristics of Reversible Circuits for Faster Combinational Equivalence Checking*, submitted to DATE'16

Index

© Springer International Publishing Switzerland 2017
L.G. Amaru, *New Data Structures and Algorithms for Logic
Synthesis and Verification*, DOI 10.1007/978-3-319-43174-1

Printed in the United States
By Bookmasters